No problem can stand the assault of sustained thinking. — **Voltaire**

Problems are like washing machines. They twist us, spin us and knock us around but in the end we come out cleaner, brighter and better than before.
—*Anonymous*

Do the math!

theoni pappas

$3x=5$

math challenges to exercise your mind

— Wide World Publishing —

Publisher:
Wide World Publishing
P.O. Box 476
San Carlos, CA 94070
web site: http://www.wideworldpublishing.com

First Printing March, 2015
ISBN: 978-1-884550-74-4

Contents

Contents

Problems continued

Contents

Problems continued

Preface

Computation problems, problems involving advanced equations, or problems that use plain old logic — all fuel new ideas and the mathematical imagination.

Problems can be provocative and can engage your curiosity. *Do the math!* covers a range of math problems. They will tantalize, challenge or amuse you, all the while honing your math skills and logical thinking.

Do the math! is designed so that if you get stuck, you can turn to the **HINT section**.

Each problem comes with an in-depth solution, found in the **SOLUTION section**.

Hopefully these problems will stimulate your brain, challenge your wits and increase your confidence. By the end of *Do the math!* you will have enhanced your mathematical tool kit.

Remember:
Imagination + patience + tenacity = solution

—Theoni Pappas

Problems

The who's who problem

Mr. Number, Mr. Piano and Mr. Painter teach math, music and art at Lane High School, but their names do not necessarily go with their disciplines. The math teacher carpools with Mr. Piano, and is the newest teacher to the school. Mr. Number has taught at the school longer than the music teacher.

Who teaches what?

What's wrong with this diagram & why?

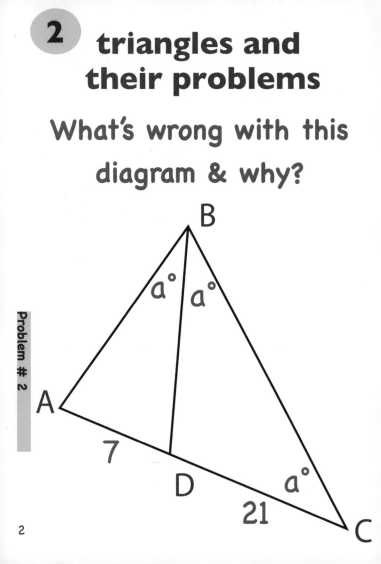

which expression  is largest?

Without using a computer or calculator determine which of these three expressions is the largest, and explain why.

$$2^{100}$$

$$3^{75}$$

$$6^{50}$$

4 What number am I?

If I am divided by 2 the remainder is 1.

If I am divided by 3 the remainder is 2.

If I am divided by 4 the remainder is 3.

If I am divided by 5 the remainder is 4.

If I am divided by 6 the remainder is 5.

What is the smallest number I can be?

?

the alphabet problem

Discover how these letters were grouped as shown, and fill in the missing letters.

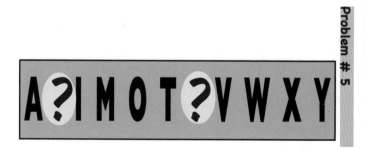

A ? I M O T ? V W X Y

license plate problem

What's the license plate number?

This car's 7 digit license plate happens to be a palindrome. Its first digit on the far left is twice the digit next to it. The 3rd digit from the right is odd and 5 more than the 2nd digit from the left. The middle digit is the average of the first three digits on the far left.

8 coins in a row problem

8 pennies are lined up in a row. Figure out a way to have these 8 pennies end up in two stacks of 4 pennies each. A penny can move only by jumping over 4 of its neighboring pennies either backwards or forwards. For example:

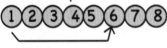

The fraction dilemma

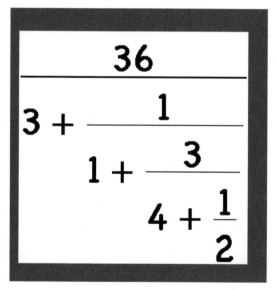

$$\cfrac{36}{3 + \cfrac{1}{1 + \cfrac{3}{4 + \cfrac{1}{2}}}}$$

What does this equal?

A square & an equilateral triangle mix it up

What is the size of the largest equilateral triangle that can be inscribed in this square?

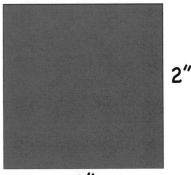

2"

2"

Connect the dots problem

Without lifting your pencil connect the 16 dots using 6 straight line segments.

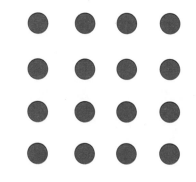

a toothpick problem

Find a way to arrange these 12 toothpicks so that you will end up with six squares with each toothpick as a side of at least one of the six squares.

The numbers come to order

Determine the answer for each problem below.

$$6 + 8 \div 2 - 2 \bullet 3 = ?$$

$$12 \div 6 \div 2 = ?$$

tetromino problems

A tetromino is a group of four identical squares connected at their sides in various ways.

The five possible tetromino shapes are:

? Which configuration(s) of the two rectangles below can or cannot be covered by a combination of tetramino shapes without overlapping or leaving gaps, and why?

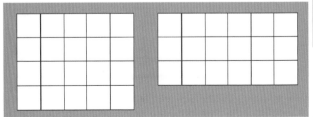

The trolley problem

The two trolley tracks, A & B, are 8 meters apart. In one trip around the tracks, trolley A goes 28mph with 8 passengers, while trolley B travels at 28mph with 12 passengers. On this trip, how many more meters does trolley A travel than trolley B?

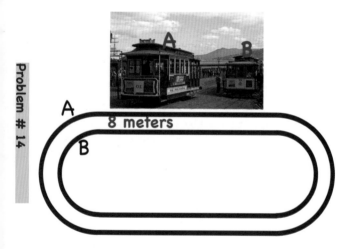

the logs problem

$$\log_2 \sqrt{\log_{10} 10,000}$$

Logarithms date back to the 17th century. In 1614 John Napier (1550-1617) was the first to publish a work on computing with logarithms and their tables, *Mirifici Logarithmorum Canonis Descriptio*. Others who worked on logarithms at this time include: Joost Bürgi(1552-1632), Henry Briggs(1536-1630) and Johannes Kepler (1571-1630).

John Napier

16

The line problem

What do these points have in common?

A is located at (1, 3)

B is located at (-2, 9)

C is located at (0, 5)

D is located at (6, -7)

stacking squares problem

Eight identical squares have been stacked as shown. Number the squares in the order they were originally placed down. The last square that was placed is number 8.

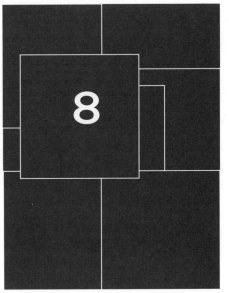

What comes next?

Figure out the pattern, and determine the next number in this sequence of numbers.

2, 6, 12, 20, 30, 42, 56, 72, 90, 110, 132, ?, 182, ...

As time goes by problem

I have both a 3 minute and a 5 minute sand timer. I am cooking something that needs to be cooked for about one minute. Find a way for me to use the timers to measure approximately 1 minute.

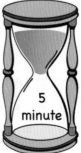

The missing digits problem

x and y represent digits that were missing from this multiplication problem. What digits do x and y have to be?

$$
\begin{array}{r}
5964 \\
\times \ \ 7xy \\
\hline
4{,}y11{,}972
\end{array}
$$

getting the operations right

Without moving around the seven digits on the left of the equal sign, figure out a way to place the operation signs +, -, ×, ÷, and parentheses to make the seven digits to the left of the "=" equal to 1.

There is no restriction on how many +, -, ×, ÷, or ( ) you can use.

Factorial problems

A
$$\frac{9!}{\left(2^3\right)!} = ?$$

B How many different ways can a deck of 52 cards be shuffled?

The term 5!, standing for 5 factorial, was once written as $\lfloor 5$.

the melting problem

Ted orders a super-duper ice cream cone. A perfect spherical scoop of chocolate ice cream is placed on a perfect isosceles circular cone whose sides are 6" & base is 3" in diameter. The ice cream scooper is also 3" in diameter.

Just as Ted is about to take a lick his cell phone rings. He has to take this important call, and holds his cell in one hand and the cone in the other. He does not dare begin eating it while talking to his boss. When he finally hangs up, he discovers his cone has melted. Did the ice cream overflow the cone, or did the cone accommodate the ice cream?

24

the six matchsticks problem

Suppose you are given six matchsticks and asked to form four triangles so that each side of each triangle is one matchstick long. How is this done?

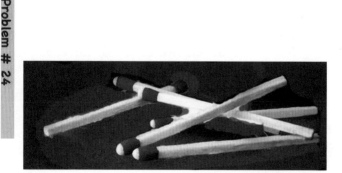

the circle problem

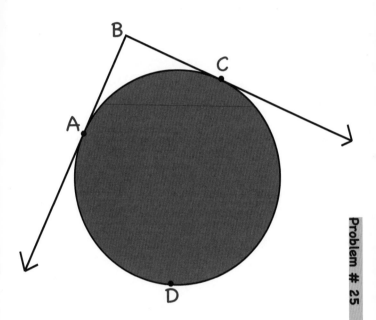

If arc ADC=270° and |AB|=5", what is the length of the circle's diameter?

the number line-up problem

Study this sequence of numbers. Which term of the sequence would **34.25** be?

2.75, 5, 7.25, 9.5, 11.75, 14,...
1st 2nd 3rd 4th 5th 6th...

34.25

Where do I belong?

the inspectors problem

Termite inspectors Tom and Jake inspected a home today with a very narrow crawl space. After 45 minutes, Tom crawled out first and Jake followed shortly. Tom's head was full of cobwebs. Jake immediately began dusting his hair with his hands. Tom watched, but did not bother with his curly dark head of hair. Can you explain both men's actions?

the card problem

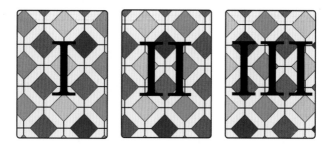

From a deck of cards, three cards are placed face down as shown. There are no face cards. Cards I and II total 12, and II and III total 16. No card is greater than 9 in value, and no card has a value of 7. What values do cards I, II, III have?

aging problem

Mr. Vain doesn't like discussing his age. So when someone asks him his age or when he was born, he replies "I mentioned in 1953 that sometime during my life my age squared equaled a particular year."

Figure out in which year Mr. Vain was born.*

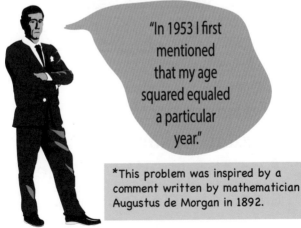

"In 1953 I first mentioned that my age squared equaled a particular year."

*This problem was inspired by a comment written by mathematician Augustus de Morgan in 1892.

Problem # 29

hexagon & triangle problem

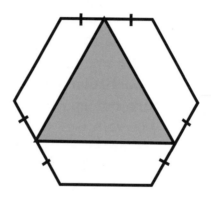

If the perimeter of this regular hexagon is 60cm, find the lengths of the sides of triangle ABC.

A coin problem

Six identical pennies are stacked as shown. Each has diameter 3/4". What is the height of the stack?

the fishy problem

The weather and time of year were perfect for trout fishing on the river. So two fathers decided to take their two sons fishing. After four hours the dads decided it was time to call it quits since everyone had caught a trout. What's the least number of trout caught?

unraveling
the trig function
problem

$$\left(\frac{\sin \theta^{\circ}}{1 + \cos \theta^{\circ}} \right) \times \left(\csc \theta^{\circ} + \cot \theta^{\circ} \right)$$

$$= ?$$

Simplify this expression, and determine for which values of θ the answer is not true.

how many pipes?

I plan to replace a 1" diameter pipe that is 10 feet long with some 1/2" diameter pipes. How many 1/2" pipes would I need to carry the same amount of water as the 1" diameter pipe along the 10 feet span?

decoding the penny problem

Study the hidden code in the coins shown, and determine the missing coins in the last statement.

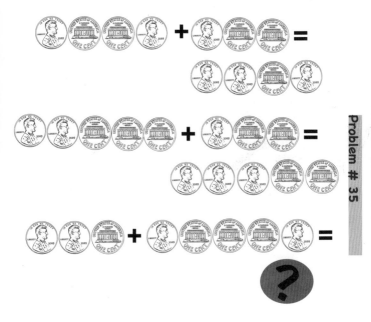

the cat on the fence problem

A 15 foot ladder is leaning against the top of a fence. Its base is 9 feet from the base of the fence. A tabby cat scurries up the top of the ladder and jumps into the neighbor's garden on the other side of the fence. The cat landed 16 feet from the fence directly in line with the ladder. If the tabby's path was a diagonal path from the top of the ladder, what was the distance of the cat's path?

divisor problem

What's the smallest number that can be divided evenly by these digits?

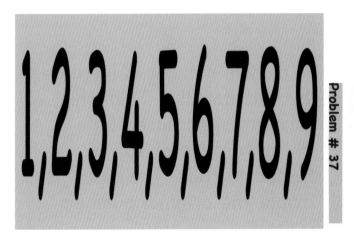

What's the number?

A certain number is increased by half its value. Then this amount is increased by a third its value and 18 is added to that. The result is triple the original number.

What's the original number?

cutting the square problem

Take a square sheet of paper. Using a pair of scissors, figure out a way to make one cut, so that the square is cut into five pieces.

a water pan problem

3 quarts

5 quarts

Mr. Cheapo had a 3 quart and a 5 quart pan. He needed to get 4 quarts of water in the 5 quart pan without having to buy a measuring cup. Figure out two different ways to this.

pile of pebbles problem

Suppose these 700 pebbles are all about the same size and shape. They can be arranged to form different shaped rectangles. What is the largest number of pebbles from this pile you can use which can only form one rectangle shape?

the sailing problem

At 5:00am a fishing boat left port heading due north for 50 miles. The fish were not biting at this location so the boat reversed its course and headed 30 miles south, and then changed its bearings and traveled 10 miles east. Still no luck here, so the captain headed 12 miles north and 38 miles west. When it arrived at this

location an alert came through directing the captain to head 13 miles north. At this location all the fisherpeople on board caught their limit in an hour, and the boat headed back to the harbor. How many miles was it back to the harbor?

the cannon ball problem

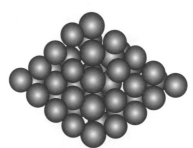

These cannon balls are stacked in the form of a regular tetrahedron with 5 balls on each side of the bottom row. **How many cannon balls are in the pile?**

If there were a regular sided tetrahedron of cannon balls with **k** balls on each side of the bottom row, how many balls (in terms of k) will there be in the bottom layer plus the layer above it?

which coin is missing?

Jeff had 12 coins on his desk. As he was transferring them to his pants pocket he accidently dropped them on the floor. He could only find 11 coins when he  gathered them up. The 11 coins were worth $1.10. Jeff had noticed just before he dropped the coins that the number of quarters plus the number of dimes was two less than the number of nickels.

Which coin denomination is missing?

The triangle card problem

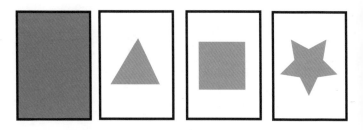

Each of these cards has a figure on only one side. Which of the four cards must one turn over to determine if every card with a triangle on one side is blue on the other side?

the melon problem

At a local Farmers Market, a farmer brought a load of melons. His first customer bought half of the load plus half a melon to enjoy while walking around. The second customer took half

of the remaining melons plus half a melon. The third customer also decided to take half of what was left plus half a melon. Now the farmer had only one melon left, so he decided to pack up his stand and save the last melon for himself.

How many melons were there in the original load?

a circle problem

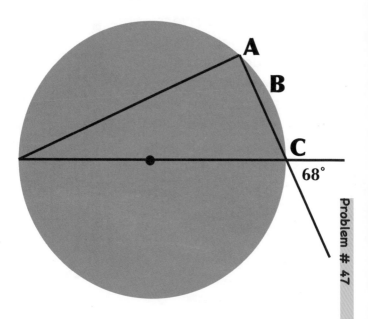

68°

measure of arc ABC=?

the tree & river problem

This tree at the river's edge is 120' tall. The river is 100' wide. Suppose the tree breaks so that its tip reaches across the river to the edge of the opposite side. Its broken parts remained attached at the break.

How many feet above the tree's trunk must the tree break?

what's the number?

3,0,0,9,7,3,0,0, 9,7,3,0,0,9,7,3, 0,0,9,7,...

In this repeating sequence of numbers, which number lands in the 3,874th place?

the word problem

Find a five letter word which when you remove two letters, only one is left.

the circle
versus
the square

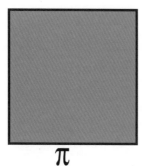

π

$\sqrt{\pi}$

Which has
the greater area,
the circle or the square?

trisecting
a triangle's angles

Draw a triangle of any shape.
Now draw in the trisecting
segments for each angle of
the triangle. Look at the
hexagon which is formed
by the trisectors.
Connect only the
vertices of the
hexagon where two
trisectors of the
Δ's adjacent
angles intersect.
**What type of triangle
is this?**

A quadrilateral's diagonals

B

3 7

5.2

7 C

3

9.6

A

8

9

8

9

D

Find the sum of the lengths of diagonals AC and BD.

54

5 heads problem

Ken tosses 5 pennies in the air. What is the probability all of the pennies will land heads up?

Problem # 54

54

the way of
the gears problem

• If gear A is revolving counterclockwise which way does gear B turn.
• If gear D makes a revolution in 27 seconds, how long does it take gear C to revolve once?

tetrahedron problem

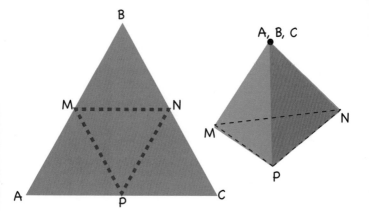

Δ ABC is an equilateral triangle with perimeter 60 units and points M, N, and P are midpoints of the triangle's sides. If the vertices A, B and C are folded along the dotted lines to form a tetrahedron, then what is the tehtrahedron's volume in cubic units.

exponent problem

$$625^{2x-3} = 125^{x+7}$$

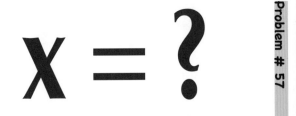

$$X = ?$$

the deli tag problem

"Take a number," yelled the deli clerk. I pulled a number tag with a 3-digit number. When I rotated the tag around it also had a 3-digit number, which was 693 less than my tag's number.

What is my tag's number?

the square & its border problem

Using a square sheet of paper, figure out a way to fold it so that when one straight cut is made with a pair of scissors a square and a border around the square result.

Is it possible to cut the original square so that the border and the interior square have the same area?

are these problems for real?

The imaginary number "i" is defined as:

$$i = \sqrt{-1}$$

Using its definition find the value of:

$$i^{100} = ?$$

$$\sqrt{i} \times ? = 1$$

the written number problem

How are twenty-two thousand, twenty-two hundred, and twenty-two written as a number?

getting a
rational answer
from an
irrational problem

What do you get when you simplify this expression?

$$\frac{9\sqrt{8} + 6\sqrt{18}}{\sqrt{50}}$$

pizza galore problem

Pizza Galore restaurant has a Tuesday Special which gives a certain number of slices of pizza to each patron

who also orders a dessert. Chuck and Sarah went on Tuesday to take advantage of the offer. They each ordered a dessert and some slices of pizza in addition to the free slices they each got. Their total order was 10 slices including the free ones, and each paid $5.25 for the pizza portion of their dinner. If Chuck had gone by himself, he would have had to pay $14 for the 10 slices of pizza. **How many free pieces does the restaurant give to each patron on Tuesdays?**

Problem # 63

63

the ancient triangle puzzle

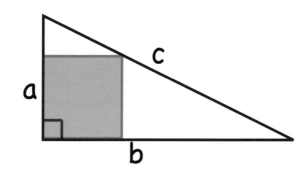

The square shown is the largest possible one that fits in the triangle's interior. Write its area in terms of **a**, **b**, and **c**.

arc & angle problem

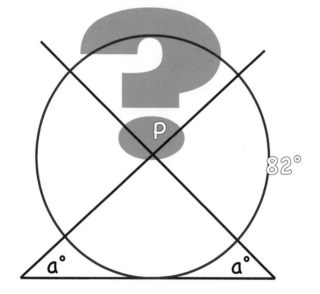

82°

a°   a°

If "P" is the circle's center, find a°.

66

the 3 ladies problem

May, Martha, and Meg were at a luncheon. Mr. Demand joined them, and immediately asked how old they were. May was taken aback by the question and blurted out 30. Martha did not like to divulge her age, and said she was May's age plus a third of Meg's age. Meg, being sensitive to May's answer, said her age was Martha's age plus a third of May's age.

How old is each?

shrinking rectangle problem

The area of this rectangle is 360 square inches. If its length is shortened by 22" and its width by 4", a square is formed.

What are the rectangle's dimensions?

the laser printer job

The small EveryReady copy shop has three laser printers of different capabilities. Printer **A** prints 5 pages in 10 seconds. Printer **B** prints 20 pages a minute. Printer **C** prints 2 pages in 20 seconds. A job just came in for 3000 copies of a single page.

If all printers are used at the same time, how many minutes will it take to print this job?

Problem # 68

how many iPhones ???

John has a box full of new iPhones to deliver to various places. To the APPLE Store he delivered half the number he had in

the box, but before John left the store, the manager checked the existing stock and decided to give back 10. From there John went to the T-MOBILE store where he delivered one-third of the remaining iPhones he had. The counter person took these except for two. His last stop was the VERIZON store where he was to leave off half of the remaining iPhones, instead he decided to keep the new red iPhone from the remaining half in order to buy it for himself. John counted the remaining inventory in his box and found there were 12 iPhones left including the red one.

Problem # 69

**How many iPhones were
originally in the box?**

the volume problem

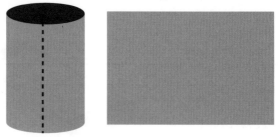

This cylinder and this rectangle are related in a special way. If the cylinder is severed along the dotted line segment and flattened, the rectangle is formed. The rectangle's length is greater than the cylinder's height. If the area of the rectangle is 27 square inches and its perimeter is 21 inches, **what is the volume of the cylinder in cubic inches?**

the hour hand problem

From 12:02pm, how many degrees would the hour hand have moved when the clock reads 1:20am the next day?

a travel problem

Tim's family lives on the outskirts of a remote town. Tim left for school on his bike going an average of 10mph. Twenty minutes later, his brother Eric realized Tim had taken his backpack rather than his own. Eric hopped in his car and traveled along Tim's route at the speed limit of 30mph, and eventually reached Tim.

How many minutes did it take Eric to reach Tim?

Marbles problem

I picked out 5 red, 9 blue and 3 yellow marbles from this box and placed them in a paperbag. How many times, without looking, must I reach in the bag and pick a marble in order to be certain I will get a red one?

the weight problem

There are four weights on this scale. The first three weigh three consecutive odd numbers and the fourth weight is the next even number after the third weight.

How much does the third weight weigh?

motorcycles problem

Two cyclists are monitored at the same moment when on opposite ends of a 3 mile long straightaway on the desert. A sound wave is continually traveling back and forth at 768mph between the emitters and sensors on each cycle. The cyclists' speeds are 66mph and 84mph. Approximately how many miles did the sound wave travel when the cyclists pass one another?

root problem

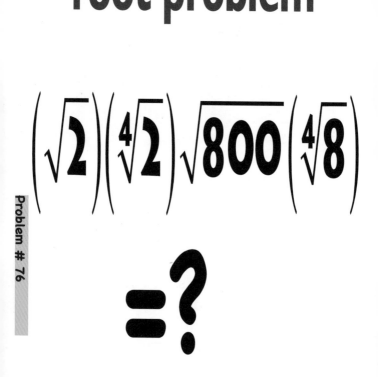

$$\left(\sqrt{2}\right)\left(\sqrt[4]{2}\right)\sqrt{800}\left(\sqrt[4]{8}\right) = \,?$$

Who watches TV?

There are 21 members in the Evans family. Thirteen watch sports. Eleven family members

watch news. Of these eleven, six only watch both the news and sports networks. Five family members watch only sitcoms and sports. Three members watch news and sitcoms. Two watch news, sports and sitcoms. Three members only watch sitcoms.

How many members do not watch TV?

At the races

Three roos were competing this year at the kangaroo races. The large roo, Biggy, jumps 10 feet with every leap. The medium sized roo, Medio, leaps 5 feet with every jump. The small roo, Tiny, always jumps in 2 feet leaps. The racetrack is a straight 42 feet long, and the race consists of a round trip. Although Medio and Tiny have to jump more often, they both have no trouble keeping up with Biggy. Medio takes two jumps for every leap Biggy makes, while Tiny takes 10 jumps.

In what order do the roos finish? And why?

Problem # 78

odd number problems

1, 3, 5, 7, 9, 11, 13, 15, 17, 19, 21, 23, 25, 27, 29, ..., 2863, 2865, 2867,...

1) If the odd numbers are listed in a sequence starting with the odd number 1, which term of the sequence is the odd number 2683?

2) What is the sum of the odd numbers from 1 through 2683?

3) What is the sum of the odd numbers from 683 through 2683?

Problem # 79

the raffle problem

A church was planning to sell raffle tickets for a 4 day cruise to the Hawaiian islands. If the tickets are priced at $5 each, the church makes $1,031. On the other hand, if they sell the tickets at $7 each, they make $1761.

How many tickets must be sold for the different profits, and what did the church pay for the cruise?

the zero problem

000 0000000 000
000000 000
0000000000
00000...

Determine how many zeros
are at the end of

125!

82

the social group problem

A social group for various activities has just completed its open enrollment. Each member signed up for one activity. Half of its members are signed up for drama, one-seventh for chess, one-fourth for dance, and three members for mathematics activities.

How many members are there in this social club?

area problem

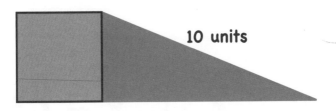

10 units

A square and a triangle are lined up as shown. The right triangle's longest leg is triple the square's side.

What's the square's area?

the intersection problem

$$y = 2^x \ \& \ y = 2x$$

At which point or points do these two equations intersect?

pattern problem

⬜	⭐	🔺	⚫	20
⬜	⬜	⚫	⚫	18
⭐	⭐	🔺	⬜	24
⬜	⬜	⬜	⬜	28

Each of the four shapes represents an integer. The numbers in the column at the right indicate the sum of the shapes' integers in that row.

Find the value of each shape.

What comes next?

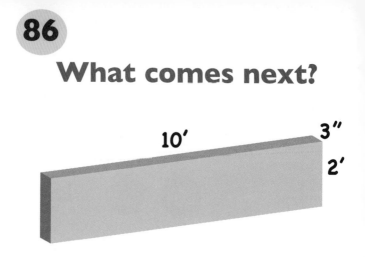

10'   3"
                2'

A carpenter has a board that is 10' long, 2' wide, and 3" thick. He wants to saw off a piece with volume 1 2/3 cubic feet.

How long will his remaining piece be?

an age problem

Two friends, Mary and Elaine, ran into each other at Starbucks on Saturday. They had not seen each other for years. Mary asked Elaine how many children she had. Elaine replied "three."

"What are their ages?" asked Mary.

"The eldest children are twins," Elaine replied, "and my youngest child is 3' 2" tall. In fact, the product of their ages is 32."

"You haven't changed Elaine," Mary said. "You still answer my questions with a problem."

How old are Elaine's three children?

88

the minute hand problem

If the minute hand of this clock moves 2" in 3 minutes, about how long is the minute hand?

the picnic problem

At the Smith's family picnic the Gather Game is always played. 80 apples are placed in a straight line with each apple two feet apart from its adjacent apples. A basket is placed 2 feet to the left of the 1st apple. Each player must start at the basket and pickup the first apple in the line, and then place it in the basket. Then, that same player goes to the 2nd apple, and returns and places it in the basket until all the apples have been picked up. Each person must follow the process of going back and forth between the apples and the basket until all the apples are gathered by each player.

The winner is the one who completes the task in the shortest time.

About how many miles does each participant have to walk before completing the task?

money problem

Four school friends got part-time summer jobs. Between the four of them they earned $114.25 the first week, so they decided to celebrate and meet at the movies. On his way, Ted collected $8 he had lent a friend. When Ned got to the movie theater he noticed he had lost $3. Ben decided to borrow some money from his father, who agreed to loan him as much as Ben made the first week. Jed, on the other hand, saw a friend on the way to the movie theater to whom he lent half his first week's earnings. When they all gathered at the movie theater and compared what each had, they all had the same amount of money.

How much had each earned at their part-time jobs that week?

The quadrant problem

II | I

III | IV

In which quadrant is the terminal side of an angle in standard position with radian measure $\dfrac{53\pi}{12}$?

92

the light problem

There is a row of 1999 button lights all in the off position.

Press the 2nd button on the left, and press every other button. The resulting pattern is shown above. Now start back at the 1st light, press every third button. Then go back and press every 4th button. Then every fifth button, and so on until you've pressed every 996th button.

Is light 996 on or off?

it went viral

Chuck wanted to test the impact of communicating via emails. He decided to email a political message to 10 friends and ask each of these friends to email it to 10 different friends, who would be asked to email it to 10 different friends, and so on. Chuck sent his email at 12pm. If each phase of the emailing took one minute for each group to email, **how many messages were sent out at 1:40pm?**

what kind of triangle?

The coodinates (2,4); (6,1) & (-1,0) represent 3 vertices of a triangle. Determine specifically what kind of triangle they form when graphed, and prove your conclusion.

the three carpenters punch in problem

Ed, Tom, and Chuck are three carpenters working on a house remodeling job. Ed always punches his time card when he arrives, Tom never does, and Chuck does half of the time. On Monday the foreman heard the first carpenter arrive and punch his time card, but he forgot to look who it was. What's the probability the next carpenter to arrive will punch his card?

infinity & square roots

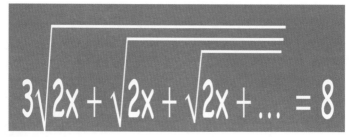

$$3\sqrt{2x + \sqrt{2x + \sqrt{2x + \ldots}}} = 8$$

Determine the value of x.

the camel run

A camel and its owner have to deliver 30 pounds of feed to a village 60 miles away. The person buying the feed also wants to buy the camel. The owner has 180 pounds of feed. The camel can only carry 60 pounds of feed at a time, and consumes one pound of feed for each mile traveled. How can the trip be arranged so that both 30 pounds of feed and the camel can be delivered to the village?*

Problem # 97

---

*This type of transportation problem dates back over 1300 years.

infinite area?

The white square's perimeter is 8 units.
The pink square's vertices are at the
midpoints of the white square's sides.
The blue square's vertices are at the
midpoints of the pink square's sides, and
so on. If this process is carried out ad
infinitum, **what is the sum of all the
areas of the squares ?**

the eccentric artist

Francois wants to paint what he calls his fuzzy masterpiece. He plans to name it the ESSENCE OF GRAY.

For this huge painting done in acrylics he calculates he will

need 10 gallons of paint which must be 4% black and 96% white. He has plenty of two other gray paints that he wants to mix to produce this new gray. One of his gray paints is 2% black and 98% white, and another 10% black and 90% white. **How much of each must he combine to end up with 10 gallons?**

Hints

Hint-1

Make a chart similar to the one shown here, and eliminate people by using the given information.

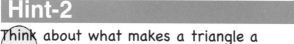

Hint-2

Think about what makes a triangle a triangle, and determine |AB|.

Hint-3

$$2^{100} = 2^{50} \cdot 2^{50}$$

Hint-4

Notice that each remainder is 1 less than its divisor.

Hint-5

Look in the mirror.

When the digits of a numeric palindrome are reversed, the number reads the same as the original number.

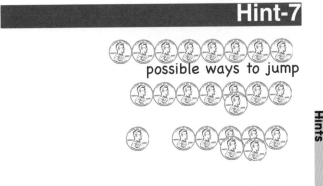

possible ways to jump

Break this problem up into multiple problems, and begin from the bottom working your way up.

Hints

Hint-9

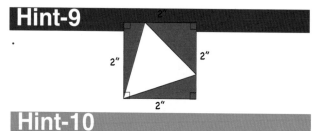

Hint-10

Think outside the box, and draw some segments beyond the dots.

Hint-11

Think on different planes.

Hint-12

When there are not any parentheses or brackets to indicate the order of the operation, the rule is: *Working from left to right, first do the multiplication and division and then the addition and subtraction also from left to right.*

Hint-13

Think about how many squares each rectangle is composed of and the factors of the number 4.

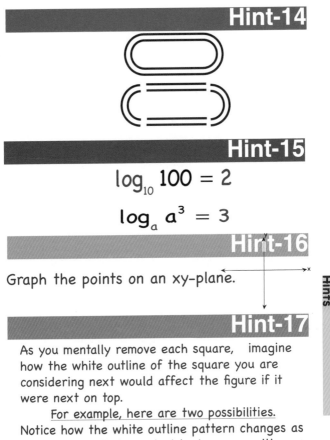

$$\log_{10} 100 = 2$$

$$\log_a a^3 = 3$$

Graph the points on an xy-plane.

As you mentally remove each square, imagine how the white outline of the square you are considering next would affect the figure if it were next on top.

<u>For example, here are two possibilities.</u>
Notice how the white outline pattern changes as the square with the asterisk changes positions.

Hints

Hint-18

Factor the numbers in the sequence in various ways to uncover a pattern.

Hint-19

Hint-20

The y in the multipler can be either a 3 or an 8 in order to produce the 2 in the product's ones place. Now find out how to produce a 7 in the tens place by looking at the possibilities for x.

Hint-21

Since we have to end up with it equaling one, consider different ways this can happen. Eventually you will end up with two numbers where either their sum, difference, product, or division equals 1. Parentheses placed in the right places will help you end up with two numbers that will work out.

Hint-22

A: 5! means
5x4x3x2x1=120

B: There are 52 possible choices for the top card

of the deck. Then 51 for the next card. Continue this process throughout the entire deck.

The volume of a sphere with radius r is:

$$\frac{4}{3}\pi r^3$$

The volume of a cone whose circular bas has radius r and height is h is:

$$\frac{1}{3}\pi r^2 \bullet h$$

Think in three dimensions

Each term increases by the same amount. Find this amount, and figure out a shortcut to get to 34.75.

Would it have made a difference if there had been a mirror?

Hints

Hint-28

Think of different combinations of cards that total 12 and those that total 16, and test which ones work out.

Hint-29

Mr. Vain's age, call it x, when squared equals a particular year (x) of his life.

Hint-30

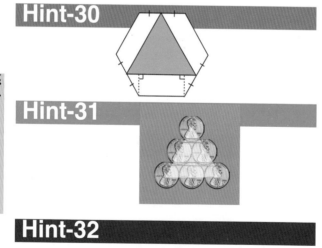

Hint-31

Hint-32

The answer is not 4 trout.

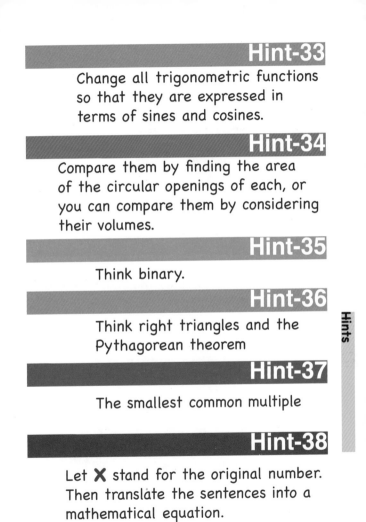

Hint-33

Change all trigonometric functions so that they are expressed in terms of sines and cosines.

Hint-34

Compare them by finding the area of the circular openings of each, or you can compare them by considering their volumes.

Hint-35

Think binary.

Hint-36

Think right triangles and the Pythagorean theorem

Hint-37

The smallest common multiple

Hint-38

Let **X** stand for the original number. Then translate the sentences into a mathematical equation.

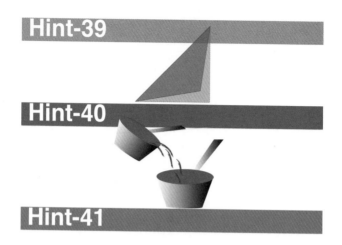

If there were 12 pebbles the shapes would be

Sketch the path of the fishing boat from the harbor.

N

50 miles

harbor

Study the number of cannon balls in each layer of the example pile.

Look for a pattern as you investigate the sum of any two consecutive layers of balls.

Think of the three possible values of the coins depending on which coin was lost. Determine which missing coin would satisfy all the conditions of the problem.

One could use trial and error to solve problem, or do it by using algebra.

Here' is one way to start with algebra:
Let n represent the number of nickels, d the number of dimes, and q the number of quarters.

Since there were originally 12 coins,
then n+d+q=12.

We also are told that q+d=n-2.

Substituting the second equation into the first we get:
n+n-2=12 —> n=7
This means **7** of the orginal twelve coins were nickels, and the remaining five coins are a mixture of dimes and quarters.

Hints

111

Hint-45

Read the problem over carefully, and think if you need to know what needs to be on the other side of each card.

Hint-46

Make a guess and work through the problem. OR, another way to attack the problem is to work backwards using algebra. For example, let x represent the number of melons left after the 2nd customer's purchase.

Hint-47

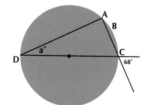

angle D is the inscribed angle of arc ABC

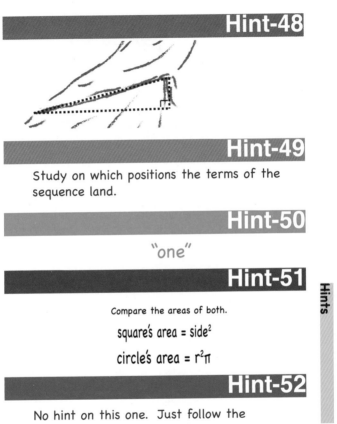

Study on which positions the terms of the sequence land.

"one"

Compare the areas of both.

square's area = side^2

circle's area = $r^2\pi$

No hint on this one. Just follow the instructions

Hints

Hint-53

Hint-54

Figure out how many different ways the pennies can land or use the Pascal triangle.

Hint-55

HINT for question 1: Look at how gear A's movement affects how C revolves. Follow your observations through until you reach gear B.

HINT for question 2: Find out how many times C revolves when D completes one revolution.

Hint-56

A, B, C

diagram 1

height

M   N

P

A tetrahedron is a pyramid, and the volume of a pyramid=(1/3)(area of its base)(height). All 4 faces of the tetrahedron are small congruent equilateral triangles with

each side 10 units. If we find the area of △ABC, then diagram 1 shows us △MNP is a fourth the area of △ABC. Now you have to figure out the tetrahedron's height.

You can also find the area of △MNP independently, since its 114 sides are all half the size of each of △ABC's sides or 10 each.

Hint-57

So long as a≠0 and $a^b=a^c$, then b=c.

Hint-58

Which digits remain digits
when they are rotated 180°?

Hint-59

Hint-60

Hint for $i^{100}=?$

$i = \sqrt{-1}$

$i^2 = -1$

$i^3 = -i$

$i^4 = i^2 \bullet i^2 = (-1)(-1) = 1$

Hint for $\sqrt{i} \times ? = 1$

$\sqrt{i} = i^{\frac{1}{2}}$

Hint-61

Add them up.

Hint-62

$$\sqrt{8} = \sqrt{2 \cdot 4}$$

Hint-63

Either use the information to write two equations involving the number of free pieces and the price per piece, or try solving the problem by guessing values.

Hint-64

Look for similar triangles.

Hint-65

Think central angle.

Hint-66

Write three equations involving the ages of the three ladies, and then solve them simultaneously.

Hint-67

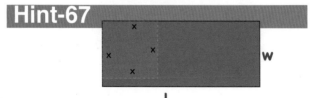

How many pages does each
printer print in a minute?

Let x represent the number of
iPhones before the last delivery,
then work backwards.

You will need to find the length
of the rectangle which will end up
also being the circumference of
the cylinder's top or base. Then
use the circumference to find the
radius of the cylinder.
**A cyclinder's volume is the area
of its base times its height.**

How many degrees does the hour
hand move in an hour?

Hints

Hint-72

$$speed \bullet time = distance$$
$$s \bullet t = d$$

Tim and Eric traveled the same distance, but traveled different speeds and lengths of time. Write a $s \bullet t = d$ equation for both Tim and Eric, and set them equal to eachother. Solve for t.

ALERT: Speeds are given in miles per hour and times in minutes.

Hint-73

Suppose the bag had only yellow and red marbles. How many times must you reach in to be certain you get a red marble?

4 times in case the first 3 you picked were yellow.

Hint-74

How many pounds apart are each of the four weights?

Hint-75

Since speed$\bullet$ time= distance.
Determine how long the motorcycles traveled before passing each other. This time interval is also how long the sound wave traveled back and forth between the cycles, which can then be used to find the distance the sound waved traveled at its speed of 768mph.

and

$$\sqrt[b]{a^c} = a^{\frac{c}{b}}$$

and

$$\sqrt[4]{2} = 2^{\frac{1}{4}}$$

Think VENN diagrams.

•Sketch the problem.

•Show the racetrack, and label the dimensions of the track and how the roos jump.

•Now think about the roundtrip.

1) HINT:
1, 3, 5, 7, ..., k, ...
terms of the sequence

1st 2nd 3rd 4th... 2k-1   Since 7=2•4-1,
places in the sequence   7 is the 4th term.

2) HINT:
  sums:
1+3=___

1+3+5=___
1+3+5+7=___ = **?²**

3) HINT: Break down this problem into two
  summation problems:
  1+3+5+...+681=? and 1+3+5+...+2863

Hints

119

Hint-80

Let x represent the number of tickets that have to be sold to get the different profits, and let c the cost paid for the cruise.

From the given information, write two equations and solve them simultaneously to find x & c.

Hint-81

Think about which numbers produce zeros when multiplied together.

Hint-82

Let x represent the total members in the activities group.

Using the information provided in the problem's description write an equation and solve it for x.

Hint-83

Hint-84

Test some values for x in each equation, or graph each equation and see where they intersect.

Study the rows and their sums. Work between rows as you determine the value of one shape to reveal the value of others.

Let x be the length of the piece of board being cut off.

List out the possible ages of the children, then eliminate those that do not meet all the criteria.

Figure out how many inches does the minute hand travels around the clock.
Recall: circumference=diameter•π

Mathematics legend, Carl Gauss (1777-1855), developed a method to total the first 100 counting numbers when he was 10 years old. Apparently his teacher assigned the class the problem of adding the first 100 counting numbers. While the rest of the class tediuosly began adding these numbers, Gauss immediately discovered an easy innovative way to do the problem. It is an approach similar to that used in the solution shown for this problem.

List the feet the player travels in the game. For example, the first apple requires the player to go 2 feet to the apple, and then two feet back to place it in the basket. List out the numbers, and look for a pattern to discover how to add them all up. How many feet in a mile?

Hint-90

We know they had all the same amount at the end, which we'll call x. Let the first initials of their names represent how much they each earned. Now follow the story, and write an equation which expresses what each earned in terms of x.

Hint-91

The red angle is in standard position. Its degree measure is 135° or (3π)/4 radians.

Hint-92

Think about when a button is pressed and the factors for that button's number, or how to tell whether there are an even or odd number of factors.

Hint-93

Figure out how many minutes passed from 12pm to 1:40pm. Now express each email-ing as powers of ten, and figure out the pattern.

Study the graph of the points. See which combination of terms best describes the triangle.
• scalene • isosceles • equilateral • equi-angular • right • obtuse • acute • isos-celes right • isosceles acute • isosceles obtuse •acute scalene   •
obtuse scalene
Use the coordinate system to prove your conclusion.

List all possible outcomes. Eliminate those that don't go along with the given information, namely that Ed always punches his time card and Chuck only sometimes, while Tom never does. This information will help decide how many outcomes for the second carpenter to arrive.

Do something to both sides of the equation, then square both sides, and continue to solve for x.

Since the camel can only carry 60 pounds of feed at a time, the trip has to be done in incremental parts. Test out whole number increments of 180 to see how the camel must travel while consuming a pound a feed per mile.

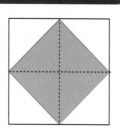

Since the black and white percentages in the given paints each total 100%, you only need to balance the black or the white percentages when combining the two given paints to produce the final 10 gallons.

Solutions

SOLUTION 1

The who's who problem

	math	music	art
Mr. Number	X	X	
Mr. Piano	X		
Mr. Painter			

The information allows us to exclude certain people from teaching certain subjects. A red x appears in the Mr.Piano/math box because we know that the math teacher carpools with Mr. Piano. So Mr. Piano cannot be the math teacher. A red x appears in the Mr. Number/math box because Mr. Number has taught the longest and the math teacher has taught the least time at the school, so they cannot be the same person. The last sentence tells us Mr. Number and the music teacher are different people, so a red x is placed in the music box in Mr. Number's row. The chart now reveals who does what. The only subject left in Mr. Number's row is art. The only person left in the math column is Mr. Painter. So that leaves Mr. Piano teaching music.

Notice △BDC is isosceles, since two of its angles are congruent, so |BD|=21.

Triangles △ABD & △ACB are similar triangles because their corresponding angles are congruent, which makes all their corresponding sides proportional.

Thus:

|AD|/|AB| = |AB|/|AC|

7/|AB| = |AB|/28

$|AB|^2 = 7 \cdot 28 = 196$

$|AB| = \sqrt{196} = 14$

But, this is not possible since the lengths of two sides of any triangle must be greater than the third side. Since |AB|+|AD|=|BD| and not |AB|+|AD|>|BD|, that diagram is impossible to construct with the given dimensions.

Solution

SOLUTION 3

Which expression is larger?

Since we can rewrite 2^{100} as $2^{50} \cdot 2^{50}$

and 6^{50} as $(3 \cdot 2)^{50}$ and then as $3^{50} \cdot 2^{50}$.

We see that $3^{50} \cdot 2^{50} > 2^{50} \cdot 2^{50}$, so

$$6^{50} > 2^{100}.$$

Now $3^{50} \cdot 2^{50}$ can also be rewritten as follows:

$$3^{50} \cdot 2^{25} \cdot 2^{25}.$$

And 3^{75} can be rewritten as $3^{50} \cdot 3^{25}$,
then we see that

$$3^{50} \cdot 2^{25} \cdot 2^{25} > 3^{50} \cdot 3^{25}, \text{ so}$$

$$6^{50} > 3^{75}$$

Therefore 6^{50} is the
largest expression.

129

4 SOLUTION

What number am I?

Since the remainders are all 1
less than the divisors, if the
answer were increased by 1,
then each of the divisors
(2,3,4,5,6) would produce a
zero remainder. So the
solution is one less than the
smallest common multiple of
the divisors, which is 60.

Thus the answer is **59**=60-1.

Solution

The missing letters are **H** and **U**. All the letters in the list are the letters from the alphabet that do not change when reflected in a mirror. They are listed alphabetically.

Solution

6 SOLUTION
The license plate problem

Since the first digit is twice the one next to it, the first digit must be an even number. The possible digits for the 2nd number are: 4, 3, 2, 1. We cannot use 5, 6, 7, 8 or 9 because twice these are not single digits.

• using 4 makes the first digit 8 and the 3rd from the right side 9

• using 3 makes the first digit 6 and the 3rd from the right side 8 — eliminate these digits because 8 is not odd.

• using 2 makes the first digit 4 and the 3rd from the right side 7

• using 1 makes the first digit 2 and the 3rd from the right side 6 — eliminate these digits because 8 is not odd. The possible plates are: 8 4 9 __ 9 4 8 or 4 2 7 __ 7 2 4

The average of the first 3 digits on the far left equals the middle digit (8+4+9)÷3=7, but (4+2+7)÷3= 13/3(which is not a digit)

This means the only answers is 8497948.

Here is one way to do the problem:

step 1 ①②③④⑤⑥⑦⑧

step 2 ②③④⑤⑥⑦⑧ (1)

step 3 ② ④⑤⑥⑦⑧ (1)(3)

step 4 ② ④⑤⑥⑦ (8)(1)(3)

step 5 ② ⑤⑥⑦ (8)(1)(3)(4)

step 6 ⑤⑥⑦ (8)(1)(3)(4)(2)

step 7 ⑤⑥⑦① (8)(1)(3)(4)(2)

step 8 ⑤⑧ ⑦①(3)(4)(6)(2)

step 9 ⑤ ⑦①(3)(6)(4)(8)(2)

step 10 ⑦①(3)(6)(4)(8)(2)(5)

The fraction dilemma

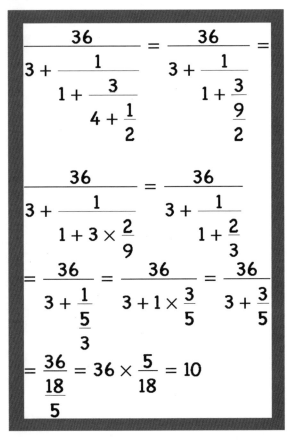

$$\cfrac{36}{3 + \cfrac{1}{1 + \cfrac{3}{4 + \frac{1}{2}}}} = \cfrac{36}{3 + \cfrac{1}{1 + \cfrac{3}{\frac{9}{2}}}} =$$

$$\cfrac{36}{3 + \cfrac{1}{1 + 3 \times \frac{2}{9}}} = \cfrac{36}{3 + \cfrac{1}{1 + \frac{2}{3}}}$$

$$= \cfrac{36}{3 + \cfrac{1}{\frac{5}{3}}} = \cfrac{36}{3 + 1 \times \frac{3}{5}} = \cfrac{36}{3 + \frac{3}{5}}$$

$$= \cfrac{36}{\frac{18}{5}} = 36 \times \frac{5}{18} = 10$$

Solution

A square & an equilateral triangle mix it up

Since the Zs in ΔABE, ΔADF and ΔECF are equal, their squares are also equal. Working with these triangles we get:

$$z^2 = z^2$$

$$2^2 + x^2 = (2-x)^2 + (2-x)^2$$

$$4 + x^2 = 4 - 4x + x^2 + 4 - 4x + x^2$$

$$0 = x^2 - 8x + 4$$

$$x = 4 \pm 2\sqrt{3}$$

$4 + 2\sqrt{3}$ is impossible because it would be longer than the square's side, thus

$$x = 4 - 2\sqrt{3}$$

Since $z^2 = 2^2 + x^2$,

$$z^2 = 4 + \left(4 - 2\sqrt{3}\right)^2$$

$$z^2 = 4 + \left(16 - 16\sqrt{3} + 12\right) - > 32 - 16\sqrt{3}$$

$$z = \sqrt{32 - 16\sqrt{3}} \approx 2.070552 \approx 2.0706 \text{ the}$$

length of the side of the equilateral Δ.

Another way to arrive at z is to note that for either ΔABE or ΔADF,

$$\frac{z}{2} = \sec 15°$$

$$z = 2 \sec 15°$$

$$z = 2\left(\frac{1}{\cos 15°}\right) \approx 2.0706$$

If none of the vertices of the equilateral triangle is on a corner of the square, then that equilateral triangle is not the largest because when one of its vertices is moved to a corner of the square (see left diagram & vertex A being moved), AE side is bigger than the original AE. So one vertex of the equilateral triangle must coincide with a corner of the square. The triangle's other two vertices will be on the sides of the square in order to be inscribed.

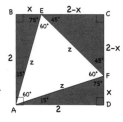

Triangles ADF & ABE are congruent because they have a pair of corresponding sides and hypotenuses congruent, which divides the right angle A in 15°, 60° and 15° angles. So the 3rd angles of these two triangles are 75° each. Since the two Δs are congruent, this makes |BE|=|FD|, which means |CE|=|CF| since the sides of the square are the same size. Therefore ΔECF is an isosceles right triangle, making its acute angles 45° each.

Apply the Pythagorean theorem to either ΔABE or ΔECF to find the actual length of hypotenuse z, which is also the length of the equilateral triangle's sides.

Solution

135

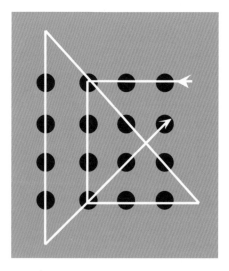

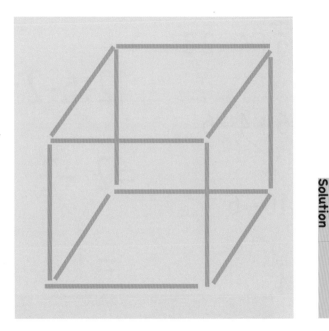

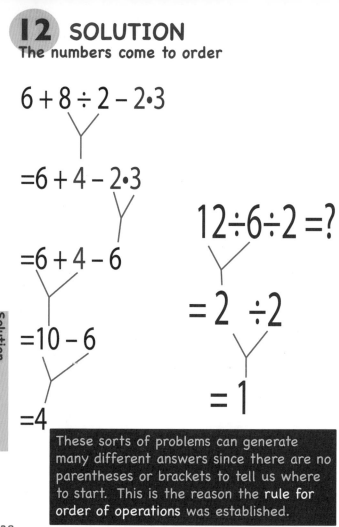

$$6 + 8 \div 2 - 2 \cdot 3$$

$$= 6 + 4 - 2 \cdot 3$$

$$= 6 + 4 - 6$$

$$= 10 - 6$$

$$= 4$$

$$12 \div 6 \div 2 = ?$$

$$= 2 \div 2$$

$$= 1$$

These sorts of problems can generate many different answers since there are no parentheses or brackets to tell us where to start. This is the reason the **rule for order of operations** was established.

Solution

Tetromino problems

The 4 by 5 rectangle is composed of 20 small
squares. Since 4 divided into 20 is 5, it can be
covered by five tetrominos. There are many
possible ways to do this. Here is one answer.

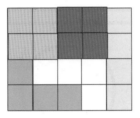

The 3 by 6 square is composed of 18 small squares.
Since 4 does not divide evenly into 18, it cannot be
covered by tetrominos. There will always be two
small squares left over.

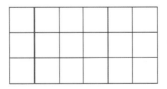

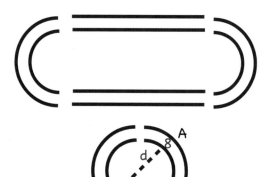

The lengths of tracks A & B only differ at their semicircular ends. So all we have to find is the difference between the circumferences of the large and small circular tracks (that is between circle A & circle B). Let d= circle's B diameter, then:

B's circumference is $d\pi$.

A's circumference is $(d+16)\pi = d\pi+16\pi$.

A's circumference-B's circumference= $d\pi+16\pi-d\pi$
   $=16\pi \approx 9.86$ meters.

the answer is 1.

$$\log_2 \sqrt{\log_{10} 10,000}$$

$$= \log_2 \sqrt{\log_{10} 10^4}$$

$$= \log_2 \sqrt{4}$$

$$= \log_2 2 = 1$$

Solution

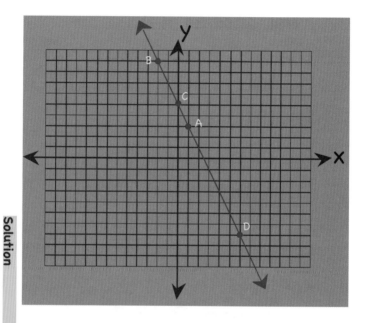

They all lie on the same line
whose equation is y=-2x+5.

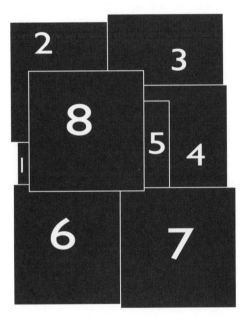

What comes next?

Factoring the numbers of the sequence in this way, reveals the pattern.

2=(1•2), 6=(2•3),
12=(3•4), 20=(4•5),
30=(5•6), 42=(6•7),
56=(7•8), 72=(8•9),
90=(9•10), 110=(10•11),
132=(11•12), ?=(12•13),
182=(13•14), ...

So ?=(12•13)=156

Turn over both the 3 and 5
minute timers at the same
time. As soon as the 3 minute
timer's sand has run out, turn
it over immediately again,
in order to measure 6 minutes.
Now, the moment the 5 minute
timer runs out, turn the 3
minute timer on its side. There
will be one minute of sand left.

Solution

The y in the multiplier can be either a 3 or an 8 in order to produce the 2 in the product's ones place, and we see that:

```
    5964              5964
  x 7x3            x  7x8
  -------          -------
  17892            44712
```

To produce the 7 in the product's tens place only a 2 or a 4 will work for x.

```
    5964              5964
  x 723            x  748
  -------          -------
  17892            44712
  11928            23856
```

Finish off the multiplication problem, and find that the multiplier is 723, making x=2 & y=3.

SOLUTION 21
Getting the operations right

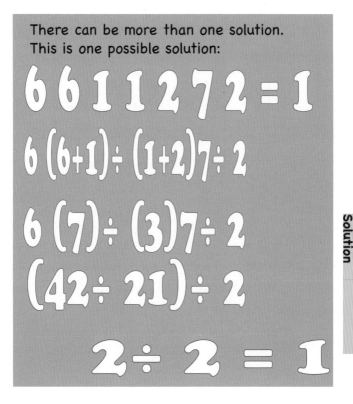

There can be more than one solution.
This is one possible solution:

$$6\ 6\ 1\ 1\ 2\ 7\ 2 = 1$$

$$6\ (6+1) \div (1+2)7 \div 2$$

$$6\ (7) \div (3)7 \div 2$$

$$(42 \div 21) \div 2$$

$$2 \div 2 = 1$$

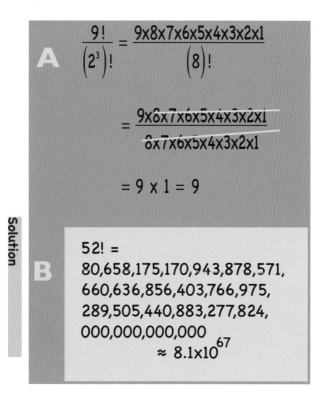

A

$$\frac{9!}{(2^3)!} = \frac{9 \times 8 \times 7 \times 6 \times 5 \times 4 \times 3 \times 2 \times 1}{(8)!}$$

$$= \frac{9 \times 8 \times 7 \times 6 \times 5 \times 4 \times 3 \times 2 \times 1}{8 \times 7 \times 6 \times 5 \times 4 \times 3 \times 2 \times 1}$$

$$= 9 \times 1 = 9$$

B

52! =
80,658,175,170,943,878,571,
660,636,856,403,766,975,
289,505,440,883,277,824,
000,000,000,000

$$\approx 8.1 \times 10^{67}$$

Solution

SOLUTION 23
The melting problem

One way to approach this problem is to find the volume of both the scoop of ice cream and the cone, and compare their sizes.

Comparing their volumes_____

Since the diameter of the scoop is 3", its radius is 1.5". So its volume is

$(4/3)\pi(1.5)^3$ =4.5π ≈ 14.14... cubic inches

The cone's radius is also 1.5". We first have to find the height, h, of the cone. Here we use the Pythagorean theorem to determine h.

$1.5^2 + h^2 = 6^2$
$h^2 = 6^2 - 1.5^2$
$h^2 = 36 - 2.25 = 33.75$
$h = \sqrt{33.75} \approx 5.81$

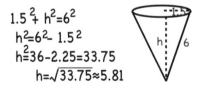

Using this value for h, the volume of the cone is:
$(1/3)\pi(1.5)^2 \cdot \sqrt{33.75}$ = $(.75)\pi\sqrt{33.75}$ ≈ 13.69 cubic in.

So the ice cream overflows.

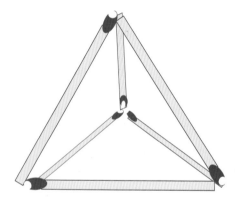

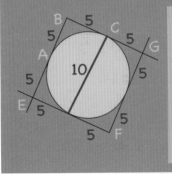

Since arc ADC is 270°, that makes arc APC=90° because the circle is 360°. So we can form a square which circumscribes the circle as shown below. Since the square circumscribes the circle, the length of the square's side will also be the circle's diameter, which equals 10".

The following Euclidean geometry theorem was used: If circle tangents AB and BC intersect, then |AB|=|BC|.

Solution

The number line-up problem

Looking at the difference between each two consecutive
terms, we find that 2.25 is
added to each term to get to
the next term. Since the sequence begins with 2.75, if we
add a certain number of 2.25's
we will get to 34.25. Or by
subtracting 2.75 from 34.25,
and dividing this result by 2.25,
we will know which term 34.25
is.

$$34.25 - 2.75 = 31.5$$
$$31.5 \div 2.25 = 14$$

Since 34.25 is the 14th term after
the first term, 34.75 is the 15th
term of the sequence.

When Tom saw Jake, he assumed he did not have anything on his head to dust off because Jake's head was clean. When Jake saw Tom's head was full of cobwebs, he assumed his also had cobwebs.

Solution

28 SOLUTION
The card problem

The possible combinations that **total 12** without using face cards are:

2,10 — this is eliminated because no card can be greater than 9.

3,9 — this is eliminated because its counterpart 9,7 has a 7.

4,8

5,7 —this is eliminated because no card can be a 7.

6,6 — this is eliminated because 6,10 was eliminated below.

The possibilities that **total 16** are:

6,10 — this is eliminated because no card can be greater than 9.

7,9 —this is eliminated because no card can be a 7.

8,8

> This leaves **4,8,8** as the values for cards I, II, III respectively.

We must find a number(Mr. Vain's age) which when squared will equal a year probably in the 20th century. If he were 10 years old that would make the year 10^2 or 100 AD, which is not near 1953. Trying other ages we find $30^2=900AD$ is still not the right century. $40^2=1600AD$ does not work. $50^2=2500AD$ is in the future. So his age must be between 40 & 50. $45^2=2025AD$, $43^2=1849AD$ is too early. $44^2=1936$. So Mr. Vain was 44 in the year 1936. Therefore his birth year was 1936-44 = **1892**.

Solution

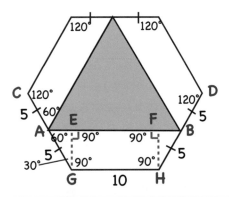

The angles of a regular hexagon are each 120°. The angles of the green triangle can be shown to be 60° each, making it an equilateral triangle. △AEG & △BFH are 30°-60°-90° triangles, so |AE| and |FB| are half of 5cm, namely 2.5cm each. So |AB|=|AE|+|EF|+|FB|. Since GEFH is a rectangle, |EF| = |GH| =10cm.

Thus, |AB|=2.5+10+2.5= 15cm

Each vertex of this equilateral triangle is at the center of a penny. So each side's length is 2 diameters. This makes each 11/2 or 1.5 inches long. The sum of lengths of the red segments is equal to the length of a penny's diameter, since they are both radii of pennies. The stack's height is the sum of the triangle's height plus the 2 red lengths.

So the penny stack's height is:
.75 + .75 + .75√3 = 1.5 + .75√3 inches, or about 2.8 inches.

Solution

3 trout.

The group consisted of a grandfather, his son, and the son's son. The grandfather and son are the two fathers, and the grandfather's son and the father's son are the two sons.

$$\left(\frac{\sin\theta^\circ}{1+\cos\theta^\circ} \right) \times \left(\csc\theta^\circ + \cot\theta^\circ \right)$$

$$= \left(\frac{\sin\theta^\circ}{1+\cos\theta^\circ} \right) \times \left(\frac{1}{\sin\theta^\circ} + \frac{\cos\theta^\circ}{\sin\theta^\circ} \right)$$

$$= \left(\frac{\sin\theta^\circ}{1+\cos\theta^\circ} \right) \times \left(\frac{1+\cos\theta^\circ}{\sin\theta^\circ} \right)$$

$$= 1$$

The expression simplies to 1, so long as θ does not produce any zeros in denominators. Studying the 3rd line from the top of the above expression we see that zeros appear in the denominators when θ is 0°, 180° or any integral multiple of 180°.

Solution

It takes **four** 1/2" diameter pipes.

Since the cyclindrical piping of the 1" diameter pipe is going to be replaced with 1/2" diameter pipes we don't need to worry about the 10' length since it is the same for all these pipes. So all we have to compare is the area of the circular bases. The formula for the area of a circle is π(radius)².

Solution

1" diameter circle's area is:

$$\pi\left(\frac{1}{2}\right)^2 = \frac{\pi}{4} \text{ sq.inches}$$

1/2" diameter circle's area is:

$$\pi\left(\frac{1}{4}\right)^2 = \frac{\pi}{16} \text{ sq.inches}$$

We see it would take four 1/2" diameter pipes.

SOLUTION 35
Decoding the penny problem

In this code the heads represent 1s and the tails represent 0s. So when the pennies patterns' are converted to 1s and 0s, the pennies write binary numbers. The first math sentence in binary numbers is 1001+100=1101, which when converted to base ten is 9+4=13. We check to see if this idea works in the 2nd example, and it does. Doing the same for the last pattern, we get:

110+10001 which in base ten is 6+17=23. 23 written in binary numbers is 10111, which converts to

heads, tails, heads, heads, heads

The cat on the fence problem

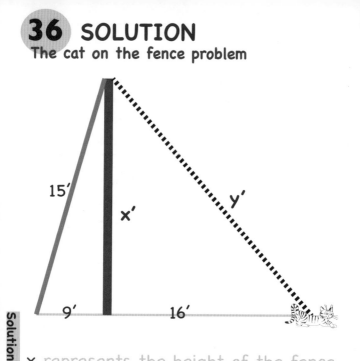

x represents the height of the fence, and y is the tabby's distance from the fence. Using the Pythagorean theorem, we get:

$$9^2 + x^2 = 15^2$$

so x=12.

$$16^2 + 12^2 = y^2$$

so y=20

So the tabby cat traveled 20′ to land 16′ from the fence.

Solution

2520

2520 contains all the prime factors for the digits from 1 through 9.

1=1•1

2=1•2

3=1•3

4=1•2•2

5=1•5

6=1•2•3

7=1•7

8=1•2•2•2

9=1•3•3

Although 1 is not a prime number, the prime factors for each digit from 1 through 9 is given.

Each digit's prime factors appear as factors of 2520.

2520=2•3•2•5•7•2•3

The first sentence asks us to combine x with half its amount, which is:

$$x + \frac{1}{2}x$$

The next sentence tells us to add a third of the above to it and then add 18 to it. So we get:

$$x + \frac{1}{2}x + \frac{1}{3}\left(x + \frac{1}{2}x\right) + 18 \quad = 3x$$

Here's one way to solve the above equation:

$$x + \frac{1}{2}x + \frac{1}{3}\left(x + \frac{1}{2}x\right) + 18 = 3x$$

$$\frac{3}{2}x + \frac{1}{3}\left(\frac{3}{2}x\right) + 18 = 3x$$

$$\frac{3}{2}x + \frac{1}{2}x + 18 = 3x$$

$$\frac{4}{2}x + 18 = 3x$$

$$2x + 18 = 3x$$

$$18 = x$$

Solution

Fold the square sheet of paper as shown in steps 1 through 3. Cut with scissors as indicated on the dotted line. You will end up with four identical triangles and a cross shaped piece.

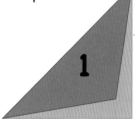

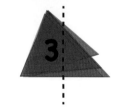

40 SOLUTION
A water pan problem

The charts below show two ways to fill and empty pans in order to end up with 4 quarts in the 5 quart pan.

- Fill up 5-pan.
- Use the 5-pan to fill up the 3-pan, which leaves 2 in the 5-pan.
- Empty the 3-pan and pour the 5-pan's 2 into the 3-pan.
- Refill 5-pan & fill 3-pan, which leaves 4 in 5-pan.

3	5
0	5
3	2
0	2
2	0
2	5
3	4

3	5
3	0
0	3
3	3
1	5
1	0
0	1
3	1
0	4

- Fill up 3-pan & pour into 5-pan.
- Refill 3-pan & pour it into 5-pan which leaves 1 in the 3-pan.
- Empty 5-pan. Pour the 1 from the 3-pan into the 5-pan.
- Refill the 3-pan and pour it into the 5-pan, which makes 4 in the 5-pan.

Solution

166

The answer is 691 because it is the largest prime number less than 700. A prime number of pebbles can only form one shaped rectangle because the only factors of any prime number are 1 and itself.

42 SOLUTION
The sailing problem

The sketch shows the path
of the fishing boat.

50 miles

N

30

13

N — W

38

12

N

E — 10 — S

20

x

32

28

harbor

Using the Pythagorean theorem:

$$28^2 + 45^2 = x^2$$
$$784 + 2025 = x^2$$
$$53 = x$$

SOLUTION 43
The cannon ball problem

Count the amount in each of the five layers, then total these.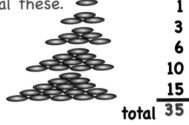

	1
	3
	6
	10
	15
total	35

Study the patterns of the sums of any of two consecutive layers above. Their sums are:

1+3=4; 3+6=9; 6+10=16 10+15=25

Notice the sums are perfect squares, which suggests the sum for layers k & k–1 would be k^2. This can be proven using the theorem that the sum of n natural numbers is

$$n+(n-1)+(n-2)+...+3+2+1 = n(n+1)/2$$

44 SOLUTION
Which coin is missing?

Continuing from the hint page

The value of the numbers of nickels, dimes, and quarters is given by the equation:
5n+10d +25q = value of the coins in cents

In the hint section we figured out **n=7**. The remaining 5 coins are dimes and quarters, so: **d+q=5** which means **q = 5-d**.
Substituting the value **7 for n** and **5-d for q** in the value equation at the top, we get:

$5 \cdot 7 + 10d + 25(5-d)$ = value of coins

This equation simplies to:

$160-15d$ = value of coins

The value of the coins found was 110¢. So the value before they were dropped could have been **115¢, 120¢,** or **135¢,** depending if the lost coin was a nickel, a dime or a quarter respectively.

Test the three possible outcomes for **d**, remembering **d** must be a whole number since it represents the number of dimes.
160-15d=115—>d=3
160-15d=120—>d=2 1/3
160-15d=135—>d=1 2/3

So we see that
the nickel was the lost coin.

You only have to turn over the card with the blue triangle on it. We are only interested if EVERY card with a triangle on one face is blue on the other side. We do not care what is on the other side of the cards with a square or a star. Whatever the blue card has on its other side is not important. Whether it has a triangle or not does not change the answer to the question. We only need to make sure that the card with a triangle is blue on its other side.

 **SOLUTION**
The melon problem

Let x represents the number of melons left
after the 2nd customer's purchase.
The equation describing the 3rd customer's
purchase is:
$x-[(1/2)x-(1/2)]=1$ solving this we get x=3.

Let y represents the number of melons left
after the 1st customer's purchase.
The equation describing the 2nd customer's
purchase is:
$y-[(1/2)y-(1/2)]=3$ solving this we get y=7.

Let z represents the number of melons in the
farmer's original load.
The equation describing the 1st customer's
purchase is:
$z-[(1/2)z-(1/2)]=7$ solving this we get z=15.

The farmer started with 15 melons.

SOLUTION 47
A circle problem

The solution of this problem depends on the geometric theorem which states that *the measure of an inscribed angle of a circle is half the size of its intercepted arc.*

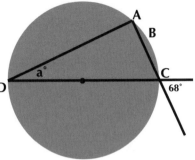

- Angle A is 90° because it is inscribed on a semicircle, which makes it half of 180°.
- Angle ACD is also 68° because it is a vertical angle of the 68° angle.
- So a° + 68°+90°=180° because the 3 angles of a triangle total 180°.
- This makes a°=22°, which means its inscribed arcABC is twice 22° or 44°.

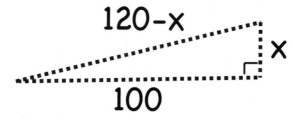

$$120-x$$

$$x$$

$$100$$

Let x represent the distance
of the break from the bottom
of the tree's trunk.

$$100^2+x^2=(120-x)^2$$

$$10,000+x^2=14,400-240x+x^2$$

$$18 \tfrac{1}{3} \text{ feet} = x$$

Look at the terms of the sequence and their respective positions:

3, 0, 2, 9, 7, 3, 0, 2, 9, 7, 3...
1 2 3 4 5 6 7 8 9 10 11,...

The sequence repeats itself in groups of five. To determine which number occupies the 3874th term, all we have to do is divide 3874 by 5. If it goes in evenly that means it is the fifth term in the cycle of 5 terms, so it is a 7. In this case, it does not go in evenly, but has a remainder of 4. So the fourth term over in this cycle of five terms is 9.

The 3,874th term is a 9.

Solution

The solution explains
itself. Some possible
solutions are:

alone
crone
drone
loner
goner
money
honey
ozone
prone
stone

Solution

In this case the areas of the square and the circle are the same.

square's area = π^2

circle's area = $(\sqrt{\pi})^2\pi = \pi^2$

52 SOLUTION
Trisecting a triangle's angles

These diagrams show trisectors from adjacent angles in the same color.

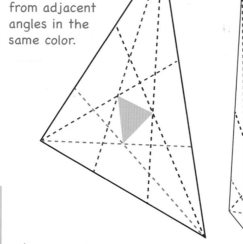

It's an equilateral triangle.

This property was first observed in 1899 by mathematician Frank Morley, and is named for him. It is interesting to learn about the various proofs of Morley's Theorem that evolved, and the number of mathematicians that were involved with it.

SOLUTION 53
A quadrilateral's diagonals

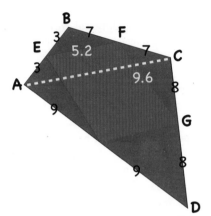

Point E and F are midpoints of the sides
of triangle ABC, which makes segment EF
parallel to AC and half its length. This
means diagonal **AC** is twice as long as EF,
namely **10.2** units.

The same reasoning follows for diagonal
BD, making its length twice that of
segment FG or **19.2** units.

Therefore the sum of the diagonals is 29.6.

Solution

179

54 SOLUTION
5 heads problem

We could start listing all the different ways the five pennies can land. For example, all heads, all tails, combinations of 4 heads and 1 tail, combinations of 3 heads and 3 tails, combinations of 2 heads and 3 tails, or combinations of 1 head and 4 tails. Once you have done this you'll find there are 32 ways with only one way having all five heads. Thus the probability of 5 pennies all landing heads up is 1/32

A short cut is to use the famous Pascal triangle numbers. One of its many properties is that its rows reveal the different ways identical coins can land. For example, its 5th row of numbers lists all the various ways 5 pennies can land. Row 4 is for a toss of 4 coins, and so on.

```
              1 row 0
            1   1 row 1
          1   2   1 row 2
        1   3   3   1 row 3
      1   4   6   4   1 row 4
    1   5  10  10   5   1 row 5
  1   6  15  20  15   6   1 row 6
  • • • • • • • • • • • • • • • • • •
```

NOTE: This is the breakdown of the combinations. Here, H stands for heads & T stands for tails.

1(5H)**+5**(4H & 1T)**+10**(3H & 2T)**+10**(2H & 3T)**+5**(1H& 4T)**+1**(5T)**=32**

HHHHH	HHHHT	HHHTT	TTTHH	TTTTH	TTTTT
	HHHTH	HHTTH	TTTHT	TTTHT	
	HHTHH	HTTHH	THHTT	TTHTT	
	HTHHH	TTHHH	HHTTT	THTTT	
	THHHH	HTHTH	THTHT	HTTTT	
		HHTHT	TTHTH		
		THHTH	HTTHT		
		THTHH	HTHTT		
		TTHHT	HTTHT		
		HTHTH	TTHTH		

180

SOLUTION 55

The way of the gears problem

PROBLEM 1: We notice that the direction of one gear causes the one it is in contact with to go the opposite direction. Therefore, gear B will be turning clockwise.

PROBLEM 2: Since gear D has 27 cogs and makes one revolution in 27 seconds, it moves one cog per second which means all the gears move one cog per second. Since gear C has 11 cogs it completes a revolution in 11 seconds.

Solution

56 SOLUTION
Tetrahedron problem

Continuing information from the HINT page:

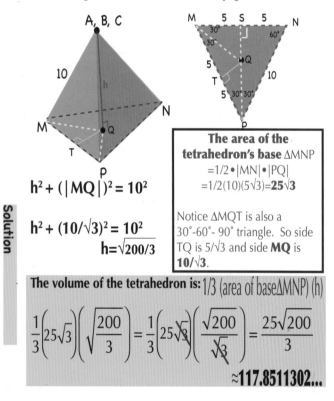

$h^2 + (|MQ|)^2 = 10^2$

$h^2 + (10/\sqrt{3})^2 = 10^2$

$h = \sqrt{200/3}$

The area of the tetrahedron's base $\triangle$MNP
$= 1/2 \bullet |MN| \bullet |PQ|$
$= 1/2(10)(5\sqrt{3}) = \mathbf{25\sqrt{3}}$

Notice $\triangle$MQT is also a 30°-60°-90° triangle. So side TQ is $5/\sqrt{3}$ and side **MQ** is **$10/\sqrt{3}$**.

The volume of the tetrahedron is: $1/3$ (area of base$\triangle$MNP) (h)

$$\frac{1}{3}\left(25\sqrt{3}\right)\left(\sqrt{\frac{200}{3}}\right) = \frac{1}{3}\left(25\sqrt{3}\right)\left(\frac{\sqrt{200}}{\sqrt{3}}\right) = \frac{25\sqrt{200}}{3}$$

$$\approx \mathbf{117.8511302...}$$

$$625^{2x-3} = 125^{x+7}$$

$$\left(5^4\right)^{2x-3} = \left(5^3\right)^{x+7}$$

$$5^{8x-12} = 5^{3x+21}$$

$$8x - 12 = 3x + 21$$

$$5x = 33$$

$$x = \frac{33}{5} = 6\frac{3}{5}$$

Solution

58 SOLUTION
The deli tag problem

We know that the number I picked and the rotated number can only be composed of the digits 0, 1, 6, 8, 9. None of the other five digits, namely 2,3,4,5,7, remain digits when rotated around. We also know that my number minus the rotated one must equal 693.

$$
\begin{array}{r}
\_\ \_\ \_ \\
\text{minus}\ \underline{\_\ \_\ \_} \\
6\ \ 9\ \ 3
\end{array}
$$

From the digits 0,1,6,8 and 9, the only pairs of two that will produce a 3 when subtracted are 9-6 or 1-8(after borrowing). So we know the ones place of my number is either 9 or 1.

The problem now becomes:

$$
\begin{array}{r}
\_\ \_\ 9 \\
-\ \underline{\_\ \_\ 6} \\
6\ 9\ 3
\end{array}
\quad \textbf{or} \quad
\begin{array}{r}
\_\ \_\ 1 \\
-\ \underline{\_\ \_\ 8} \\
6\ 9\ 3
\end{array}
$$

My number's first digit is the 100s digit on my flipped around number.

This is not a possibility because the 100s place digit 9 cannot be taken away from the 6 to produce a 6

$$
\begin{array}{r}
6\ \_\ 9 \\
-\ \underline{9\ \_\ 6} \\
6\ 9\ 3
\end{array}
\quad \textbf{or} \quad
\begin{array}{r}
8\ \_\ 1 \\
-\ \underline{1\ \_\ 8} \\
6\ 9\ 3
\end{array}
$$

The only digit for the 10s place digits to produce a 9 is if it is 1.

811-118=693

So my deli number was 811.

Solution

The square & its border problem

Fold marks will appear along the diagonals of the square.

How to make the border's area equal to the inner square's area.

To determine the width, **x**, of the border we write an equation so that both the border's and the inner square's areas are equal. They would each have to be equal to half the area of the square, which is $(1/2) s^2$, if **s** is the length of the square's side.

Let **d** be the diagonal's length and **x** the width of the border.

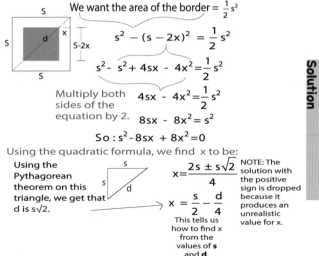

We want the area of the border = $\frac{1}{2}s^2$

$$s^2 - (s-2x)^2 = \frac{1}{2}s^2$$

$$s^2 - s^2 + 4sx - 4x^2 = \frac{1}{2}s^2$$

Multiply both sides of the equation by 2.

$$4sx - 4x^2 = \frac{1}{2}s^2$$

$$8sx - 8x^2 = s^2$$

So : $s^2 - 8sx + 8x^2 = 0$

Using the quadratic formula, we find x to be:

Using the Pythagorean theorem on this triangle, we get that d is $s\sqrt{2}$.

$$x = \frac{2s \pm s\sqrt{2}}{4}$$

$$x = \frac{s}{2} - \frac{d}{4}$$

This tells us how to find x from the values of **s** and **d**.

NOTE: The solution with the positive sign is dropped because it produces an unrealistic value for x.

Solution

Are these problems for real?

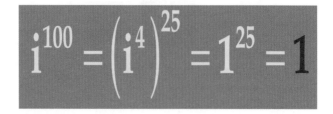

$$i^{100} = \left(i^4\right)^{25} = 1^{25} = 1$$

Here are two ways to solve the 2nd problem giving two equivalent answers.

Solution

$\sqrt{i} \times \mathbf{?} = 1$	$i^{\frac{1}{2}} \times \mathbf{?} = 1$
$i^{\frac{1}{2}} \times \mathbf{?} = 1$	$i^{\frac{1}{2}} \times \mathbf{?} = i^4$
$\mathbf{?} = \dfrac{1}{i^{\frac{1}{2}}} = i^{-\frac{1}{2}}$	$\mathbf{?} = \dfrac{i^4}{i^{\frac{1}{2}}}$
	$\mathbf{?} = i^{\frac{7}{2}}$

Adding twenty-two thousand, twenty-two hundred, and twenty two we get:

$$
\begin{array}{r}
22,000 \\
2,200 \\
+22 \\
\hline
24,222
\end{array}
$$

Solution

62 SOLUTION
Getting a rational answer from an irrational problem

$$\frac{9\sqrt{8}+6\sqrt{18}}{\sqrt{50}}$$

$$=\frac{9\sqrt{2\cdot 4}+6\sqrt{2\cdot 9}}{\sqrt{2\cdot 25}}$$

$$=\frac{9\sqrt{2}\sqrt{4}+6\sqrt{2}\sqrt{9}}{\sqrt{2}\sqrt{25}}$$

$$=\frac{\sqrt{2}\left(9\sqrt{4}+6\sqrt{9}\right)}{\sqrt{2}\sqrt{25}}$$

$$=\frac{9\cdot 2+6\cdot 3}{5}=\frac{36}{5}=7\frac{1}{5}$$

Solution

188

SOLUTION 63
Pizza galore problem

The answer is 2 slices of free pizza.

Let x represent the number of free slices given per customer. So Chuck and Sarah each got x free slices. Let p stand for the price per slice after the x free slices.

The equation for what Chuck and Sarah paid is: $(10-2x)p = \$10.50$
The equation for what Chuck would pay if he went alone is: $(10-x)p = \$14$

Solving these simultaneously we get:
$(10-2x)p = \$10.50 \longrightarrow$ **$10p-2xp = \$10.50$**
$(10-x)p = \$14 \longrightarrow 10p-xp = \$14 \longrightarrow$ **$20p-2xp = \$28$**

$$
\begin{array}{r}
10p-2xp=10.50 \\
- \quad 20p-2xp=28 \\
\hline
-10p=-17.50 \longrightarrow p=\$1.75
\end{array}
$$

Substituting this for p in one of the original equations, we get:
$(10-x)\,1.75=14 \longrightarrow$ **x≈2 free slices**

Another way to solve the problem:
The difference of the cost is $3.50 ($14-$10.50=$3.50) results from the fact that Sarah got x free slices with her dessert. Thus, if p is the price per slice, then $xp=\$3.50$.

We also know Chuck went alone and got 10 pieces for $14, so $10p-xp=\$14$.

$$xp=\$3.50$$

Solve these two equations simulatneously.

Using substitution we get $10p=\$17.50 \longrightarrow p=\$1.75$

Since $xp=\$3.50 \longrightarrow \$1.75x=\$3.50 \longrightarrow$ x≈2.

Solution

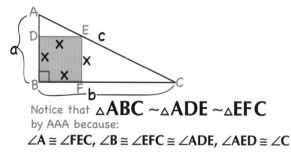

Notice that $\triangle ABC \sim \triangle ADE \sim \triangle EFC$
by AAA because:
$\angle A \cong \angle FEC, \angle B \cong \angle EFC \cong \angle ADE, \angle AED \cong \angle C$

Therefore, AB/BC =AD/DE —>

$$\frac{a}{b} = \frac{(a - x)}{x}$$

$$ax = ab - bx$$

$$ax - bx = ab$$

$$x(a + b) = ab$$

$$x = \frac{ab}{a + b}$$

$$\boxed{\text{area of the square is } \left(\frac{ab}{a + b}\right)^2}$$

Solution

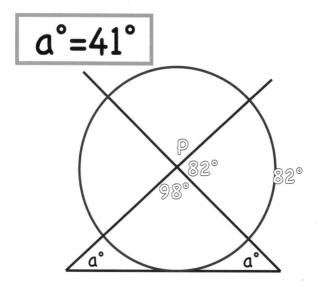

$$a° = 41°$$

You can find either a° by considering that the three angles of a triangle total 180°: a°+a°+98=180°
OR consider that the exterior angle of the triangle is the sum of the two remote interior angles:
a°+a°=82°—>a°=41°.

Solution

66 SOLUTION
The 3 ladies problem

A represents May's age, **B** is for Martha's and C is for Meg's age.

May said she was 30, so **A=30**

Martha said she was May's ages plus 1/3 of Meg's: **B=A+(1/3)C —> B=30+(1/3)C**

Meg said her age is Martha's plus 1/3 of May's:

$C=B+(1/3)A$ ⟹ **C=B+(1/3)30—>C=B+10**

B=30+(1/3)C
B=30+(1/3)(B+10)
B=30+(1/3)B+10/3
(2/3)B=30+10/3
B=50, which makes C=50+10 or C=60.

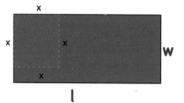

The area of a rectangle is length times width, so $l \cdot w = 360$.

Since the length was shortened by 22, we know $l - 22 = x$.

Since the width was shortened by 4, we also know that $w - 4 = x$. Since these two statements equal x, we can write:

$l - 22 = w - 4$, which means $l = w + 18$. Substituting this in for l above:

$l \cdot w = 360$

$(w + 18) \cdot w = 360$

$w^2 + 18w = 360$

$w^2 + 18w - 360 = 0$

$(w - 12)(w + 30) = 0$

w=12" or $w = -30$ (which we eliminate since width is a positive quantity.)

Since $l \cdot w = 360$, **l=30"**

68 SOLUTION
The laser printer job

Since it takes printer A ten seconds to print 5 pages, then in one minute it prints 30 pages.

We are told printer B prints 20 pages in a minute.

Since printer C prints 2 pages in 20 seconds, it completes 6 pages in a minute.

Working together they print 30+20+6 pages in a minute or 56 pages a minute.

So the 3000 pages takes
3000/56≈ 53.57 minutes
working together.

Solution

If **x** represents the number of iPhones before the last delivery to the VERIZON store, then half of these plus the red one equals 12.

$$(1/2)x+1=12$$
x=22

Now let y represent the number of iPhones before the T-Mobile delivery, so

$$(1/3)y+2=22$$
y=30

Finally, let z represent the number of iPhones before the APPLE Store delivery, then

$$(1/2)z+10=30$$
z=40 iPhones

which was the number John originally had in his delivery box.

Solution

Let ℓ represent the length of the
rectangle and w its width.
The rectangle's area is:
$\ell \bullet w = 27 \rightarrow w = 27/\ell$

The rectangle's perimeter is:
$2\ell + 2w = 21$
$2\ell + 2(27/\ell) = 21$
$2\ell + 54/\ell = 21 \rightarrow 2\ell^2 - 21\ell + 54 = 0$
$(2\ell - 9)(\ell - 6) = 0$

So $\ell = 4.5$ or $\ell = 6$.
If $\ell = 6$, then $w = 27/6$ or **4.5.**
We ignore the solution $\ell = 4.5$ because
the length is given as greater than this
cylinder's height.

Since ℓ is 6, then $C = d \bullet \pi \rightarrow 6 = d \bullet \pi$, so
diameter $d = 6/\pi$. This makes the radius,
r, equal to $3/\pi$.

The cylinder's volume = area of base • height→
$= \pi r^2 \bullet 4.5 = \pi (3/\pi)^2 \bullet (4.5) = (9)(4.5)/\pi \approx 12.891...$cubic in.

SOLUTION 71
The hour hand problem

ANSWER: 399°

From 12:02pm to 12:02am the hour hand makes one complete revolution around the circular face of the clock, which is 360°. From 12:02am to 1:20am, the hour hand moves one hour and 18 minutes. Since the face of a clock is divided into 12 one-hour increments, an hour increment is 30° because 360°/12=30° To find out how many degrees 18 minutes is we can use ratios. 1 hour/30° is 60 minutes/30°.

$$\frac{60 \text{minutes}}{30°} = \frac{18 \text{minutes}}{x°}$$
$$x° = 9°$$

Its total degree movement from 12:02pm to 1:20am is 360°+30°+9°=399°

Solution

72 SOLUTION
A travel problem

answer: 10 minutes

Suppose we let **t** stand for the time Tim bikes until Eric reaches him. Eric's time driving would have to be twenty minutes (which is 20/60 or 1/3 hour) less than Tim's or **t-1/3** because he left twenty minutes later. We had to change 20 minutes to a fraction of an hour because the speeds are given in miles per hour. Tim & Eric traveled the same distance since they took the same route. So we can set their respective s•t=distance equations equal to one and other.

Tim's is: **10•t=d**

Eric's is: **30(t-1/3)=d**

$$10 \bullet t = 30(t-1/3)$$
$$10t = 30t - 10$$
$$10t - 30t = -10$$
$$t = 10/20 = 1/2 \text{ hour or 30 minutes}$$

Eric's time driving is **t-1/3**

or 30-20 = 10 minutes

Solution

13 times.

Since 9 blue and 3 yellow
total 12, you have to pick
13 times to be certain
the first 12 were not
blue or yellow.

Solution

Consecutive odd numbers are two units apart. An odd number and the even number after it are just one unit apart.

Let x represent the first odd number. Then:

the 1st odd is x

the 2nd odd is x+2

the 3rd odd is x+4

The 4th weight is even or 1 pound more: the 4th is x+5.

We know these total 95 pounds, so

4x+11=95 —> x=21. The heaviest odd weight is **25 pounds**.

Solution

Since the cyclists' times began and ended at the same time, the two cyclists traveled the same length of time. Let the letter t represent this time. The total of their respective distances is 3 miles.

So $66t + 84t = 3$ miles

$150t = 3$

$t = 1/50$ of an hour

This means that the sound wave also traveled 1/50 of an hour since it started and ended its travel with the cyclists. Thus, the waves traveled:

$(1/50)768 = 768/50 = \mathbf{15.36}$ **miles**

Solution

Root problem

the answer is **80**.

$$\left(\sqrt{2}\right)\left(\sqrt[4]{2}\right)\sqrt{800}\left(\sqrt[4]{8}\right)$$

$$= 2^{\frac{1}{2}} \cdot 2^{\frac{1}{4}} \cdot \sqrt{100 \cdot 8} \cdot \sqrt[4]{2^3}$$

$$= 2^{\frac{1}{2}} \cdot 2^{\frac{1}{4}} \cdot \sqrt{100 \cdot 8} \cdot 2^{\frac{3}{4}}$$

$$= 2^{\frac{1}{2}} \cdot 2^{\frac{1}{4}} \cdot 10\sqrt{8} \cdot 2^{\frac{3}{4}}$$

$$= 2^{\frac{1}{2}} \cdot 2^{\frac{1}{4}} \cdot 10\sqrt{2^3} \cdot 2^{\frac{3}{4}}$$

$$= 2^{\frac{1}{2}} \cdot 2^{\frac{1}{4}} \cdot 10\left(2^{\frac{3}{2}}\right) \cdot 2^{\frac{3}{4}}$$

$$= 2^{\frac{4}{2}} \cdot 2^{1} \cdot 10$$

$$= 4 \cdot 2 \cdot 10 = 80$$

Solution

SOLUTION 77
Who watches TV?

One way to organize the information for this problem is by using a Venn diagram. Using a Venn diagram, make three circles. One for **news**, one for **sports**, and one for **sitcoms**.

Since we know two members watch **all three** networks, we place the number 2 in the space of the three overlapping circles.

Since six watch **only** news and sports, we place the number 6 where **only** the news and sports circles overlap.

Since five member **only** watch sitcoms and sports, we place the number 5 where **only** the sitcoms and sports circles overlap.

Since three members watch news and sitcoms (but not only those), we see that **two are** already in this region so we only have to add **1** in that part of the region where only news and sitcoms overlap, and thereby **making the total for the entire news/sitcom region 3**.

Since 11 members watch news and so far the news circle has a total of 9 members, **2** needs be added to the only watch news region.

Since three members only watch sitcoms, **3** is added to the only sitcom region.

Looking at the sports circle we see we have 13 sports watchers, so that checks out with the second sentence of the problem.

Totaling the numbers in all circles tells us that 19 members watch TV. Since there are 21 family members, **2 do not watch TV**.

Tiny comes in first, then
Medio, and finally Biggy.

Why? At the 42 foot
marker, Biggy's jump lands
him at 50 feet, Medio's
lands him at 45 feet, while
Tiny lands exactly at 42
feet, thereby giving Tiny a
head start on the return
trip.

Solution

1) 2863=2k-1 —> 2684=2k —> k=1342 or 2863 is the **1342nd term**

2) Hint 2 implies that the sum of k odd terms is k^2. Since 2863 is the 1342nd term, the sum of the odd numbers through 2863 is 1342^2 or **1,800,964**.

3) The sum of **683+685+...+2863** can be broken down to:

1+3+5+...+2863 minus 1+3+5+...+681=**683+685+...+2863**

1+3+5+...+2863=1,800,964 and
1+3+5+...+681= 340^2 = 115,600 because 681 is the 340th in the series.

$$1,800,964 - 115,600 = \mathbf{1,685,364}$$

Solution

80 SOLUTION
The raffle problem

Let x represent the number of tickets and c represents the cost paid by the church.

Since the value of the tickets **equals** the cost of the cruise plus the profit, we can write the following equations:

$$5x=c+1031 \ \& \ 7x=c+1761$$

solving simultaneously we get:

$$7x=c+1761$$
$$\text{minus} \quad 5x=c+1031$$
$$\overline{\quad\quad 2x=730 \quad\quad}$$

$$x=365 \text{ tickets}$$

Substitute x into either of the red equations to find c. For example,

$$5(365)=c+1031 \rightarrow$$

$$c=\$794 \text{ cost of cruise}$$

SOLUTION  81
The zero problem

5 multiplied by any even number will produce a zero at the end. SO to find out how many zeros will be at the end of 125!, we just need to count how many factors of 5 there would be in the numbers appearing in:

$$1 \bullet 2 \bullet 3 \bullet ... \bullet 123 \bullet 124 \bullet 125 = 125!.$$

All numbers ending in 0 will have a 5. These are 10,20,30,40,50(has 2 factors of 5),60,70,80,90,100(has 2 factors of 5), 110, 120. This total 14 fives.

Then we include all numbers ending in five. These are 5, 15, 25(has 2 factors of 5), 35,45,55,65,75(has 2 factors of 5), 85,95,105,115, 125(has 3 factors of 5) **for a total of** 17 fives.

So 125! ends in 31 zeros.

Let x represent the total number of members in the activities group.

x is composed of 1/2 drama, 1/7 chess, 1/4 dance and 3 math members.

In equation form this information is:

$$\frac{1}{2}x + \frac{1}{7}x + \frac{1}{4}x + 3 = x$$

Solving:

$$28\left(\frac{1}{2}x + \frac{1}{7}x + \frac{1}{4}x + 3 = x\right)$$

$$14x + 4x + 7x + 84 = 28x$$

$$84 = 28x - 25x$$

$$84 = 3x$$

$$28 = x$$

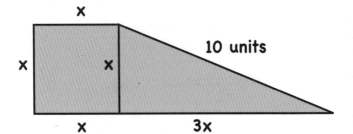

Using the Pythagorean theorem we find x as follows:

$x^2+(3x)^2=10^2$

$10x^2=100$

$x^2=10$

$x=\sqrt{10}$

The area of the square is x^2 or $\sqrt{10}^2 =10$ square units.

Solution

84 SOLUTION
The intersection problem

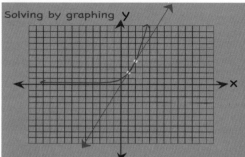

Solving by graphing

Graphing $y=2^x$ & the line $y=2x$, we see they intersect at **(1, 2)** & **(2, 4)**. We can check and verify these by substituting the coordinates in both equations:
$2=2^1$ & $2=2\bullet 1$ and $4=2^2$ & $4=2\bullet 2$

Solving by algebra

Set $y=2^x$ & $y=2x$ equal to eachother, since they both equal y. With $2^x=2x$, we get:

$$2^x = 2x$$

$$\frac{2^x}{2} = x$$

$$2^{x-1} = x$$

Now take the log of both sides:

$$\log 2^{x-1} = \log x$$

$$(x-1)\log 2 = \log x$$

$$x\log 2 - \log 2 = \log x$$

$$x\log 2 = \log x + \log 2$$

$$x = \frac{\log x}{\log 2} + 1$$

Trying different values for x, we find that x=1 is a solution since log1 is zero. The other solution is x=2, since it also produces a true statement.

$$2 = \frac{\log 2}{\log 2} + 1$$

$$2 = 2$$

Solution

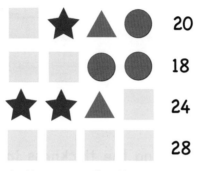

1) The bottom row of yellow squares tells us its value is 7. (square=7)

2) Replace the square's value in row two, and we see that the orange circles are each 2. (circle=2)

3) Placing the square's and circle's value in row one, we find that the star and the triangle total 11.

4) Finally, in row three by replacing the square's value and the combo value of the star & triangle, we determine that star's value is 6, which makes the triangle's value 5. (star=6) (triangle=5)

Solution

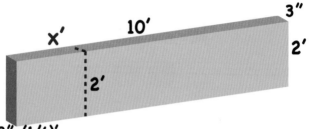

3"
10'
x'
2'
2'
3"=(1/4)'

After converting the thickness to feet, the volume is:

$$\left(\frac{1}{4}\right)(2)(x) = 1\frac{2}{3}$$

$$\frac{x}{2} = \frac{5}{3}$$

$$x = \frac{10}{3} = 3\frac{1}{3} \text{ feet}$$

Solution

The remaining board's length is:

$$10 - 3\frac{1}{3} = 6\frac{2}{3} \text{ feet}$$

Since the products of their ages is 32, the possible ages are:

$$1 \quad 1 \quad 32$$
$$1 \quad 2 \quad 16$$
$$1 \quad 4 \quad 8$$
$$2 \quad 2 \quad 8$$
$$2 \quad 4 \quad 4$$

Since Elaine has twins, that eliminates all except (2,2,8) and (2,4,4). We also know her youngest is not a twin, since she used the word youngest. So,

her children's ages are: 2,4, and 4.

Solution

Every three minutes the minute hand travels 2", so in 60 minutes or once around the clock, it would have traveled 2"(60÷3) =40". So the circumference of the minute hand is 40". We know **C=d•π** where C is the circumference and d is the diameter. Substituting 40" for C, we get:

$$40"=d•π$$
$$40/π =d$$
$$12.73323... ≈ d$$

So the radius is the minute hand's length, or
(12.73323...)÷2 ≈ 6.366...' .

Listing out the number of feet each player must traverse, we get:

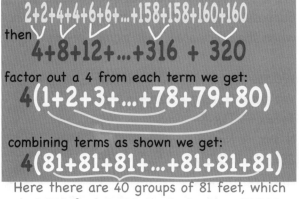

then
$$2+2+4+4+6+6+...+158+158+160+160$$
$$4+8+12+...+316 + 320$$

factor out a 4 from each term we get:
$$4(1+2+3+...+78+79+80)$$

combining terms as shown we get:
$$4(81+81+81+...+81+81+81)$$

Here there are 40 groups of 81 feet, which is 3,240 feet. 3,240 times **4** makes 12,960 feet. Since there are 5280 feet in a mile, the player travels:

$$12,960 \div 5280 = 2.454545...\text{miles}$$

Solution

90 SOLUTION
Money problem

Let T stand for what Ted made.
N for Ned's earnings
B for Ben's earnings
J for Jed's earnings
We know together their earnings totaled $114.25.
Suppose x represents the amount they each had
when they compared their monies at the theater.

We see that T+N+B+J = $114.25
We know from the scenario that
T+8=x → T=x-8
N-3=x → N=x+3
2B=x → B=(1/2)x
(1/2)J=x → J=2x

$$T+N+B+J = \$114.25$$
$$(x-8)+(x+3)+[(1/2)x]+(2x) = \$114.25$$
$$(9/2)x-5= \$114.25$$
$$(9/2)x=\$119.25$$
$$x=\$26.50$$

Replacing this up above for x, we get
T=$18.50; N=$29.50; B=$13.25; J=$53

Solution

SOLUTION 91
The quadrant problem

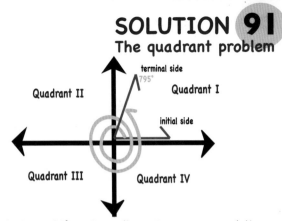

A circle has 360° or 2π radians. So once around the axes is 360° or 2π radians. Since 360°=2π radians, then 180° = π radians, or each radian is 180°/π. If we change 53π/12 radians to degrees as follows:

$$\frac{53\pi}{12} \bullet \frac{180°}{\pi} = \frac{9540°}{12} = 795°$$

Then, 795°÷360° tells us that a 795°makes two complete revolutions around the axes plus 75°, which makes the terminal side of the angle land in quadrant I.

Another way to approach the problem is to write 53π/12 as the mixed number (4 5/12)π.
The 4π tells us that the angle made two revolutions, since each revolution is 2π radians.
Since **180° = π radians**, we can convert 5π/12 radians to degrees. We get 5π/12 = 5/12 • 180° = **75°**, which means the terminal side lands in quardrant I.

92 SOLUTION
The light problem

Here are two ways to approach this problem:

Method (1) A light's button is pressed when its button numbers are 2, 3, 4, 5, 6,..., or 996,.... For example, consider button 15. It gets pressed with numbers 3, 5, and 15. Since it started in the "off" position, it is pressed with its factor 3 to the "on" position, then pressed again with the factor 5 back to "off", and finally back to "on" with the factor 15.

So we need to find all the different factors of 996 after 1 because each of these will be a time when the 996th button is pressed. These are: **2, 3, 4, 6, 12, 83, 166, 249, 332, 498, 996.**

We know it was "on" with **2**, and alternates "off" & "on" with the other factors.

So 996 ends in the "on" position.

Method (2) It is not necessary to identify a number's specific factors. You only need to know if there are an odd or even numbers of factors after 1. If there are an odd number of factors the light will be "off", and if an even number of factors it will be "on". Notice that all numbers, except perfect squares, have factors that come in distinct pairs. For example, the pair factors for the perfect 16 are (1,16); (2,8); (4,4). When its factors are listed the 4 is not repeated. They are 1,2,4,8,16. It has 4 (or an even) number of factors other after 1. **Since 996 is not a perfect square, it has an odd number of unique factors after 1. So its light is "on".** So all perfect square numbers will end up "off", and non-perfect square numbers will end up "on".

SOLUTION 93
It went viral

The 1st emailing went to 10 people at 12pm.

The 2nd emailing at 12:01pm would be 10^2 emails.

The 3rd emailing at 12:03pm would be 10^3 emails.

The 4th emailing at 12:04pm would be 10^4 emails.

From 12pm to 1:40pm is 100 minutes.

The 100th emailing at 1:40pm would be 10^{100} emails, which equal 1 followed by 100 zeros. Theoretically a googol* of messages went out at 1:40pm.

Historical note:

*The math term googol first appeared in print in the book MATHEMATICS & THE IMAGINATION by Ed Kasner & James Newman. The term was coined by Kasner's nine year old nephew in 1939. The book also introduced googol's counterpart, googolplex which equals $googol^{googol} = 10^{10^{100}}$

Solution

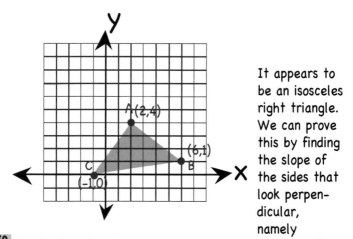

It appears to be an isosceles right triangle. We can prove this by finding the slope of the sides that look perpendicular, namely

AC & AB. If their slopes are negative reciprocals, that proves they are perpendicular, and therefore form a right angle.

The slope of AC is (4-0)/(2-(-1))=4/3.   negative

The slope of AB is (4-1)/(2-6)=3/-4.   reciprocals

To show it is isosceles we must prove the lengths of sides AC & AB are equal.

$$|AC| = \sqrt{(2-(-1))^2 + (4-0)^2} = \sqrt{3^2 + 4^2} = \sqrt{25} = 5$$

$$|AB| = \sqrt{(2-6)^2 + (4-1)^2} = \sqrt{(-4)^2 + 3^2} = \sqrt{25} = 5$$

Therefore, this is an isosceles right triangle.

Solution

Let E stand for Ed, T for Tom, and C for Chuck. Let N stand for "no punched card" and P for "punched card". So the possibilities for the carpenters are: EP represents Ed who always punches his card, TN represents Tim who never punches his card, CP means Chuck when he punched his card, and CN means Chuck when he did not punch his card.

The possible outcomes for the first carpenter arriving are: EP, TN, CP, CN. Of these we can eliminate TN and CN because the foreman heard the card being punched.

The possible outcomes for:

1ST	2ND	3RD CARPENTER
EP,	CP,	TN
EP,	CN,	TN
EP,	TN,	CP
EP,	TN,	CN
CP,	EP,	TN
CP,	TN,	EP

There are a total of 6 possible outcomes, in which the first carpenter has punched his card. Of these, there are only 2 possibilities that show the second carpenter also punching his card. These are: EP, CP, TN and CP, EP, TN. This makes the **probability that the 2nd carpenter will punch his card equal to 2/6 = 1/3.**

Solution

Inifinity and square roots

$$3\sqrt{2x + \sqrt{2x + \sqrt{2x + \ldots}}} = 8$$

$$= \frac{8}{3}$$

Squaring both sides of the equation, we get:

$$2x + \sqrt{2x + \sqrt{2x + \sqrt{2x + \ldots}}} = \frac{64}{9}$$

Now replace $\sqrt{2x + \sqrt{2x + \sqrt{2x + \ldots}}}$ with $\frac{8}{3}$.

We get:

$$2x + \frac{8}{3} = \frac{64}{9}$$

$$2x = \frac{64}{9} - \frac{8}{3}$$

$$2x = \frac{40}{9}$$

$$x = \frac{20}{9} = 2\frac{2}{9}$$

We know the incremental stops into which we divide the 180 miles cannot exceed 60 miles because the camel can only carry 60 pounds at a time. We also know it has to be much less than that because it has to have feed to get back to pick up the next 60 pounds of feed. For example, trying a 30 miles increment with 60 pounds of feed only allows the camel to make a 30 mile round trip with no feed left over. The job can be done with 15 mile increments as follows:

First increment of trip: The camel goes 15 miles, and the owner drops off 30 pounds at a campsite. The owner uses the remaining 15 pounds to feed the camel on the trip home.

Second increment of trip: The next day the camel goes 15 miles to the campsite, and having eaten 15 pounds of feed to get there leaves 45 pounds on its back. Now the owner loads 15 pounds from the 30 pounds which had been left at this campsite from the first visit. The camel proceeds 15 more miles with his load of 60 pounds. When arriving at this new campsite the camel has eaten 15 of the load, leaving 45 pounds. The owner leaves 15 pounds of the feed here, and returns home with the camel which consumes the remaining 30 pounds of its load.

Third increment of trip: The camel and the owner make their final trip with the last 60 pounds of feed. The owner stops at the first 15 mile camp, and the owner loads up the 15 pounds of feed left there. This replenishes the load to 60 pounds of feed. They now proceed to the next campsite where the owner picks up the 15 pounds of feed left there, and again replenishes the load to 60 pounds. Finally they travel the last 30 miles to the village where 30 pounds of feed remain on the camel's load, and the owner delivers this feed to his client, who happens to also buy the camel.

Solution

This diagram shows how the white square's area is half that of the pink one. Looking at the diagram below, we see how the same holds true for the pink and blue squares, and all subsequent squares within the white square. Thus each square's area is half the size of the square around it. Because the perimeter of the white square is 8 units, that makes its sides each 2 units which makes its area 2^2 or sq.4 units. The pink square's area is half this or 2. The blue square's area is half 2 or 1 sq.unit, and so on. The sum of all the areas of these squares is represented by the mathematical series shown at the top of the right column.

$$4 + 2 + 1 + \frac{1}{2} + \frac{1}{4} + \frac{1}{8} + \ldots$$
$$= 7 + \frac{1}{2} + \frac{1}{4} + \frac{1}{8} + \ldots$$

$$\frac{1}{2} + \frac{1}{4} + \frac{1}{8} + \ldots$$

Here are two ways this yellow expression can be shown to equal 1. The **first way** is by observation.

1/2	1/4	1/8	1/32
			1/16

0   1

Looking at this 1 unit long ruler, we see how the summation of the fractions gets closer & closer to 1. In calculus "1" is referred to as the series limit as it approaches infinity.

$$\frac{1}{2} + \frac{1}{4} + \frac{1}{8} + \ldots \rightarrow 1, \text{ thus}$$

$$4 + 2 + 1 + \frac{1}{2} + \frac{1}{4} + \frac{1}{8} + \ldots = 8$$

Another way to solve the problem is recognizing the expression as a geometric series with a limit. Those familiar with calculus would notice this is a geometric series with initial term 4 and ratio 1/2. Summation of such a series is given by the expression: $\dfrac{a}{1-r}$

where "a" is the initial term and "r" is the series ratio of any two consecutive terms. Substituting in 4 for a and 1/2 for r, we get:

$$\frac{a}{1-r} = \frac{4}{1 - \frac{1}{2}} = \frac{4}{\frac{1}{2}} = 8$$

Call the supply of the paint that is 2% black & 98% white paint **A**.
Call the other supply that is 10% black & 90% white paint **B**. If we mix these two so that the black percentage is 4%, this automatically makes the white 96% because together they must total 100%.

Let **x** represent how many gallons of **A**, so **10-x** will be how many gallons of **B**, since they must total 10 gallons in all. The equation below balances the amount of the blacks' percentages to combine to make the 10 gallons of 4% black paint.

$$x(2\%) + (10-x)(10\%) = 10(4\%)$$

Solving this we get:

$$0.02x + 1 - 0.1x = 0.4$$
$$-0.08x = -0.6$$
$$x = \textbf{7.5 gallons of A}$$
$$\& \ 10-7.5 = \textbf{2.5 gallons of B}$$

Solution

—About the Author—

Theoni Pappas is passionate about mathematics. A native Californian, Pappas received her B.A. from the University of California at Berkeley in 1966 and her M.A. from Stanford University in 1967. She taught high school and college mathematics for nearly two decades, then turned to writing a remarkable series of innovative books which reflect her commitment to demystifying mathematics and making the subject more approachable. Through her pithy, non-threatening and easily comprehensible style, she breaks down mathematical prejudices and barriers to help one realize that mathematics is a dynamic world of fascinating ideas that can be easily accessible to the layperson.

Her over 18 books and calendars appeal to both young and adult audiences and intrigue the "I hate math people" as well as math enthusiasts. Three of her books have been Book-of-the Month Club™ selections, and her *Joy of Mathematics* was selected as a Pick of the Paperbacks. Her books have been translated into Japanese, Finnish, French, Slovakian, Czech, Korean, Turkish, Russian, Thai, simplified and traditional Chinese, Portuguese, Italian, Vietnamese, and Spanish.

In 2000 Pappas received the Excellence in Achievement Award from the University of California Alumni Association.

In addition to mathematics, Pappas enjoys the outdoors. Her other interests include watercolor painting, photography, music, cooking and gardening.